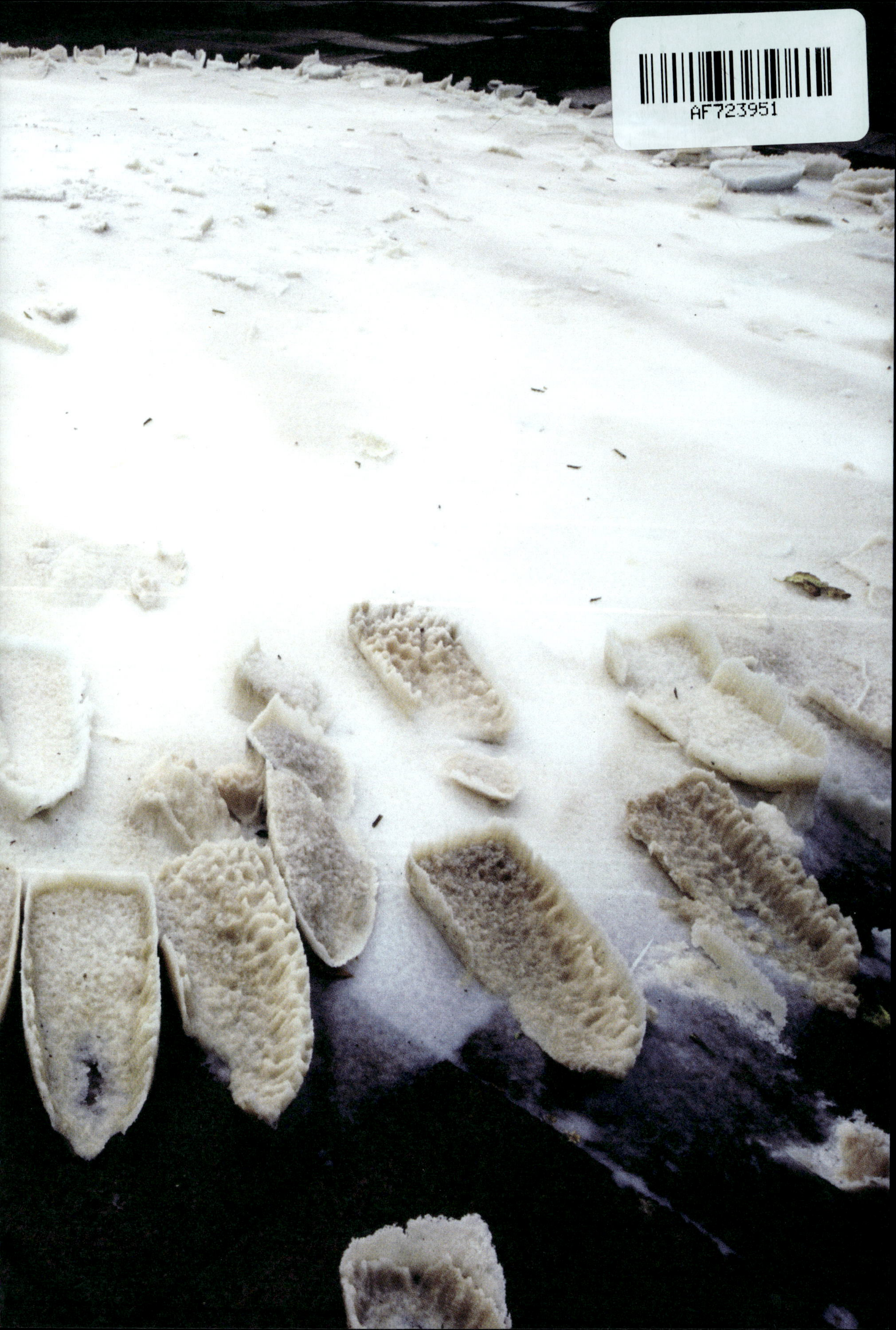

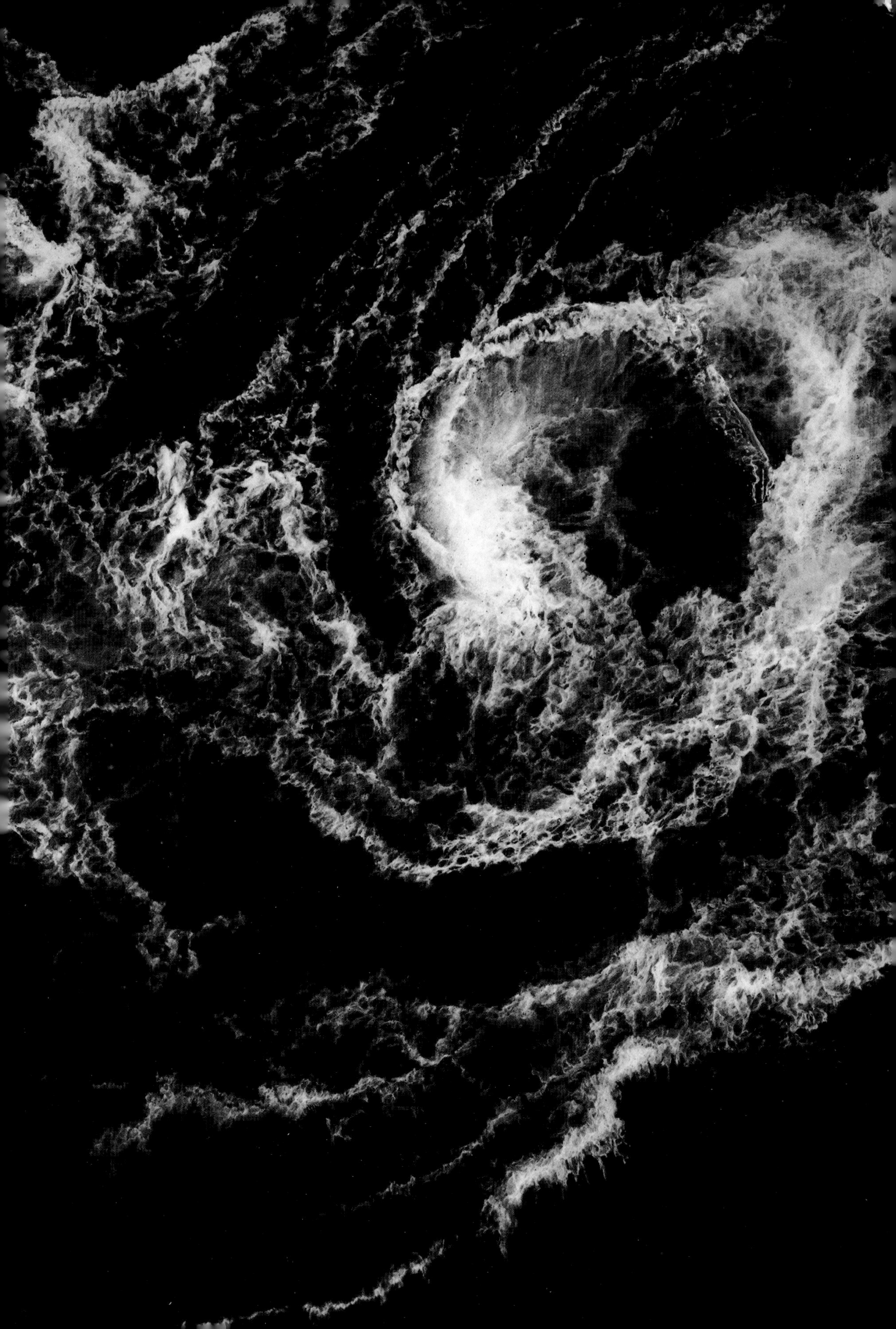

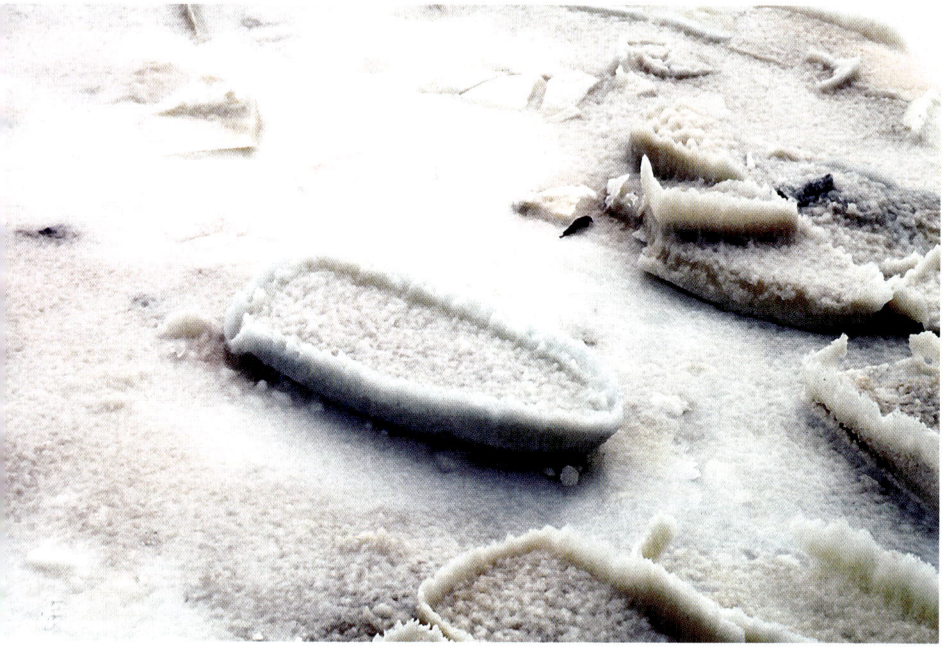

Dedicated to the memory of Yuko Yamamoto

Return to the Sea:

Saltworks by
Motoi Yamamoto

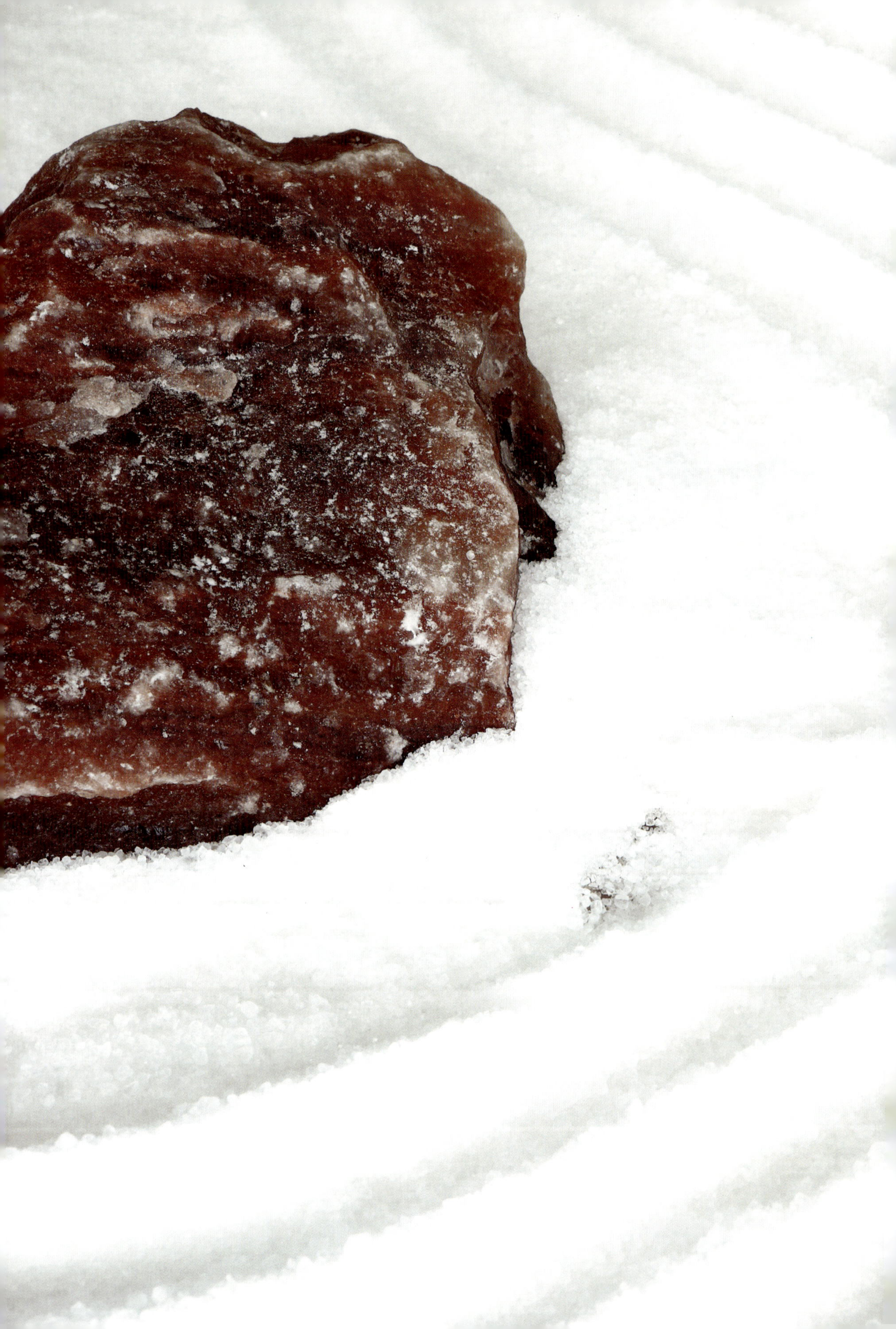

This publication accompanies the exhibition
Return to the Sea: Saltworks by Motoi Yamamoto
Originated by the Halsey Institute of Contemporary Art at the College of Charleston, South Carolina

Published by: Halsey Institute of Contemporary Art
School of the Arts
College of Charleston
161 Calhoun Street
Charleston, South Carolina 29401
843-953-4422
www.halsey.cofc.edu

Produced by the Publishing Department
Halsey Institute of Contemporary Art

Editor: Mark Sloan
Essayist: Mark Kurlansky
Archival image consultant: Stacey Clarkson
Graphic designer: gil shuler graphic design, gsgd.com
Copy editor: Gerald Zeigerman
Printing: Four Color Print Group

ISBN: 978-1-4675-2127-7

front endpaper
FOUNTAIN OF REMEMBRANCE
salt / 4 x 4m
Garden of the Sculptures, Veracruz
Veracruz, Mexico, 2000

title page
LABYRINTH
salt / 3.2 x 11m
Foundation Espace Ecureuil pour l'art Contemporain
Toulouse, France, 2011

this page
FOREST OF THIS WORLD
(detail) / 17 x 9m
salt, rock salt
The Hakone Open-Air Museum
Kanagawa, Japan, 2011
Photo: Makoto Morisawa

Contents page
LABYRINTH
salt / 3.2 x11m
Van Every/Smith Gallery
Davidson College
Davidson, North Carolina, 2006

Contents

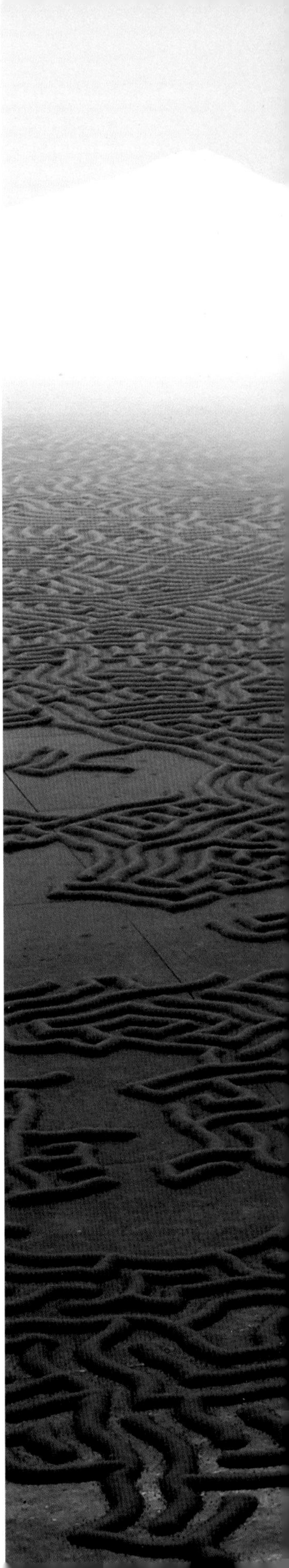

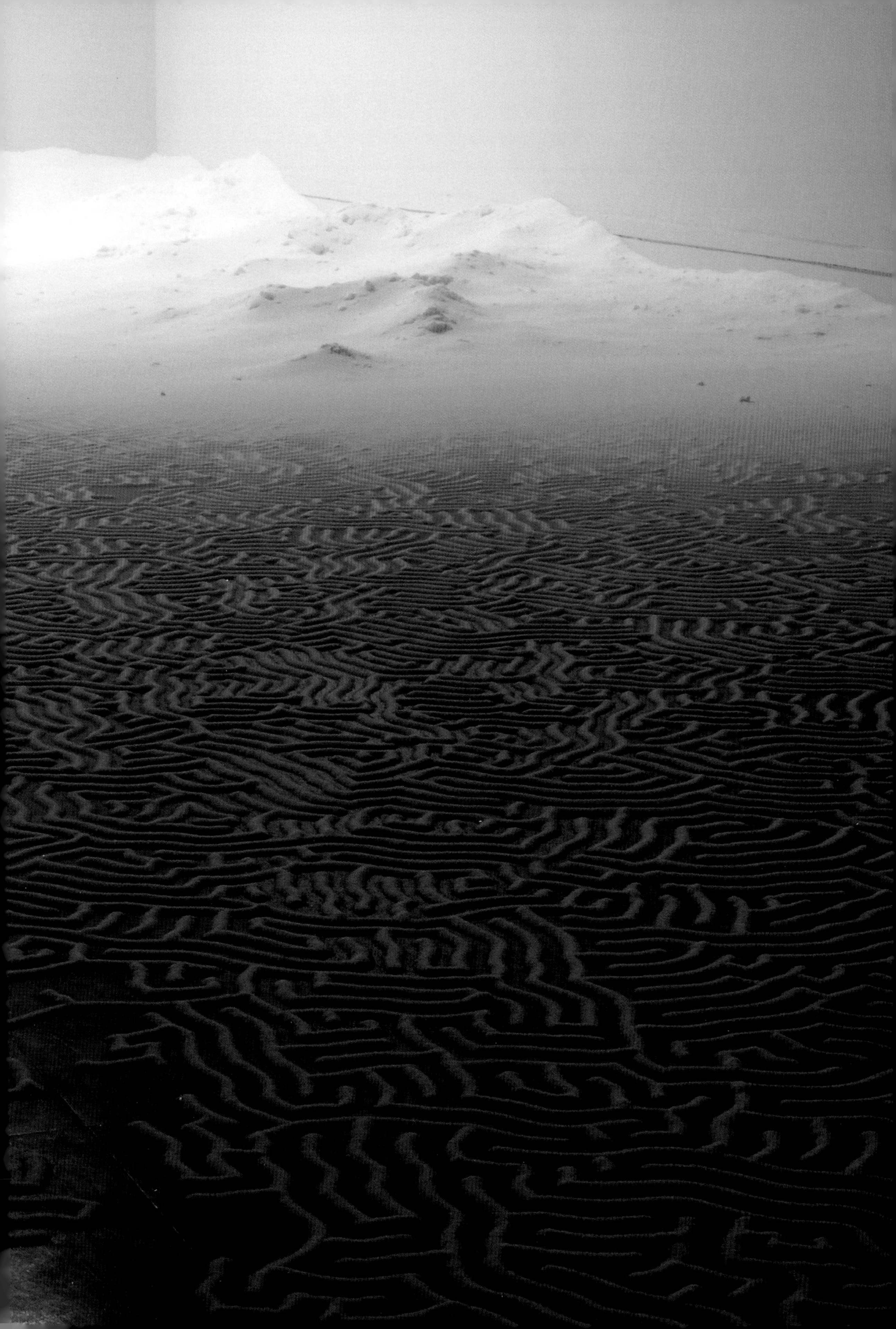

A Story in Salt

By Mark Sloan
Director and Senior Curator
Halsey Institute of Contemporary Art

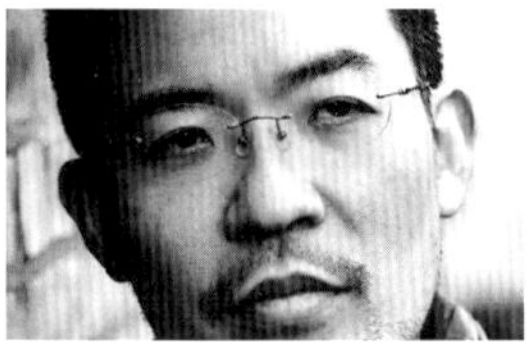

Drawing a labyrinth with salt is like following a trace of my memory. Memories seem to change and vanish as time goes by; however, what I seek is to capture a frozen moment that cannot be attained through pictures or writings. What I look for at the end of the act of drawing could be a feeling of touching a precious memory. | Motoi Yamamoto

Japanese artist Motoi Yamamoto is something of a magician. Although he does not utilize sleight of hand, he creates installations that encourage the mind to travel between astonishment and wonder. His medium is salt, yet the effect of his labors is transcendence. Not an easy feat.

The field of modern and contemporary art is crowded with artists who have worked with unconventional materials. From Meret Oppenheim's mink-lined teacup to Joseph Beuys's felt and suet, to Wolfgang Laib's use of bee pollen, the list is endless. Enter Motoi Yamamoto. He uses salt to create mental maps, miniatures of the mind. Yet, in his case, he doesn't seem to choose materials merely for the sake of novelty or originality.

Motoi Yamamoto is known for working with salt, often in the form of temporary, intricate, large-scale installations. He has created projects throughout the world—Jerusalem, New York, Tokyo, Mexico City, Athens, Seoul, Toulouse, Hamburg, and Charleston, to name but a few places. Salt, a traditional symbol for purification and mourning in Japanese culture, is employed in funeral rituals as well as by sumo wrestlers before matches. It is frequently placed in small piles at the entrance to restaurants and other businesses to ward off evil spirits and attract benevolent ones. Motoi forged a connection to the element while mourning the death of his sister at twenty-four from brain cancer and began to create art out of salt in an effort to preserve his memories. His art radiates an intense beauty and tranquility, but also conveys something ineffable, yet endless.

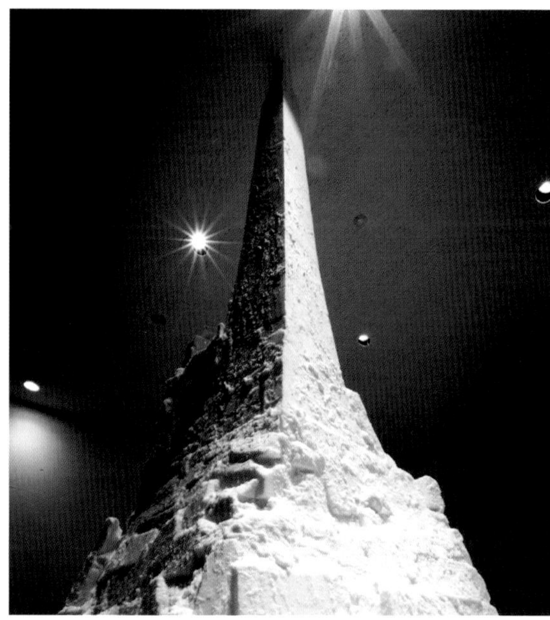

FOREST OF THE SKYSCRAPER
salt / 3.5 x 2.6 x 3.6m
The Hakone Open-Air Museum
Kanagawa, Japan, 2011

Motoi views his installations as exercises that are at once futile yet necessary to his healing. An important aspect of the installation is the dismantling of his work at the end of each show and delivering the salt back to water, usually in collaboration with the public; hence the title *Return to the Sea*.

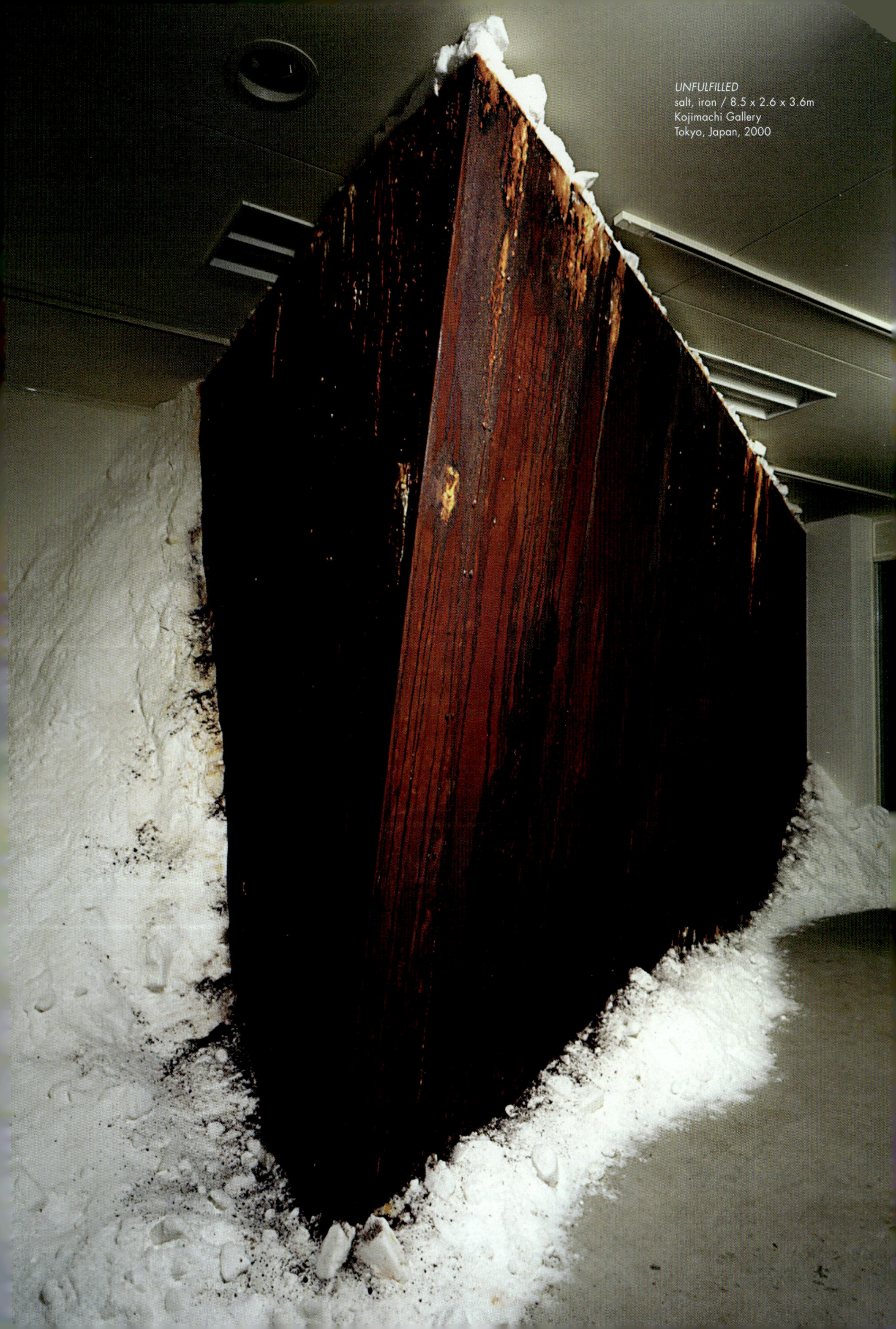

UNFULFILLED
salt, iron / 8.5 x 2.6 x 3.6m
Kojimachi Gallery
Tokyo, Japan, 2000

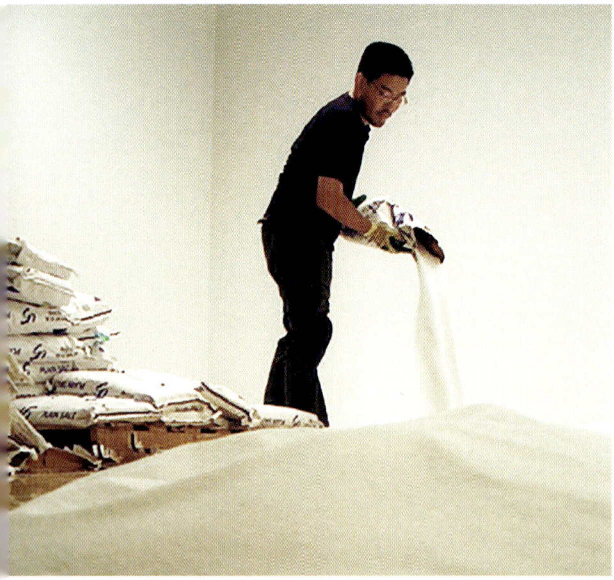

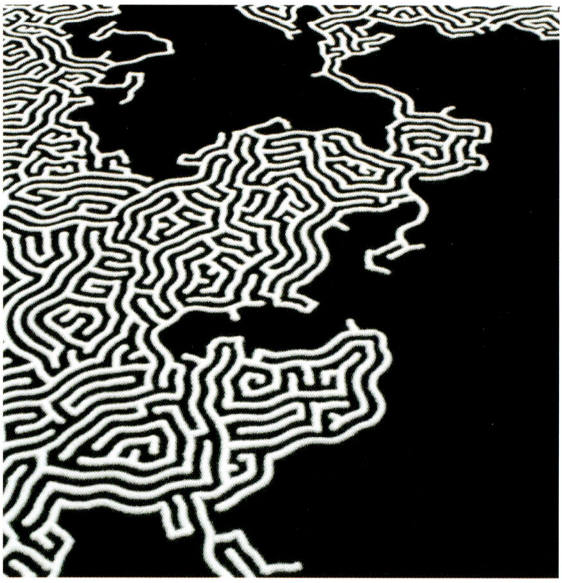

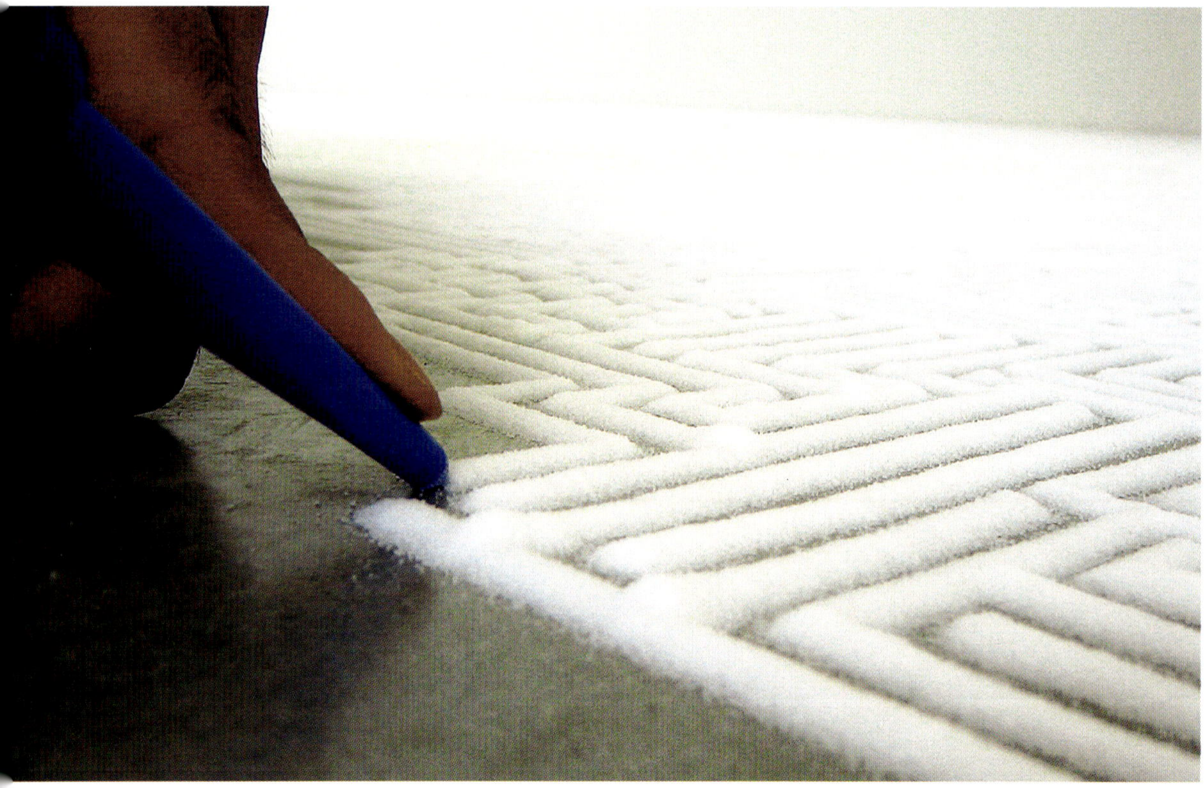

He recognizes that salt is a vital part of many living things, and that this mineral could conceivably enter and leave multiple organisms throughout the planet over the span of time. Each grain of salt contains its own history and trajectory. Something so seemingly common becomes a metaphor for the evanescence and transience of human life.

Yamamoto's art might most productively be compared to the intricate sand mandalas created by Tibetan Buddhist monks. In both cases, the work is destroyed at the end of a predetermined interval and returned to a body of water, thus enacting the circularity and ephemerality of life. Both are used to induce meditation and to access the deeper reaches of human consciousness. One important difference is that Yamamoto's work emanates from a powerful personal experience rather than a shared spiritual tradition.

This book is the first to take measure of Motoi Yamamoto's engagement with salt installations since he began making them in 1994. By showing the range of works he has produced, we seek to chart the artist's journey of healing as he explores these different forms of expression. As Mark Kurlansky states in his essay, salt has a very rich and noble history entwined with Japan's development as a world power. Knowing its myriad uses in Japanese culture makes it less of a common, everyday substance and more of a mercurial one. Its snow-white purity, combined with the uniformity of the grains, provides Yamamoto with a material at once literal and poetic—loaded with associational possibilities.

Yamamoto's artistic trajectory is full of innovations and surprises. His ability to adapt his concepts to the various configurations and idiosyncracies of the galleries and exhibition spaces that have displayed his work serves as a source of inspiration. His subtle use of gradation and perspective may cloud the mind into thinking this could be a mountain range as seen from the air or a typhoon out in the open ocean. The real power and magic of Motoi Yamamoto's work resides in the indeterminate space between what is and what might be.

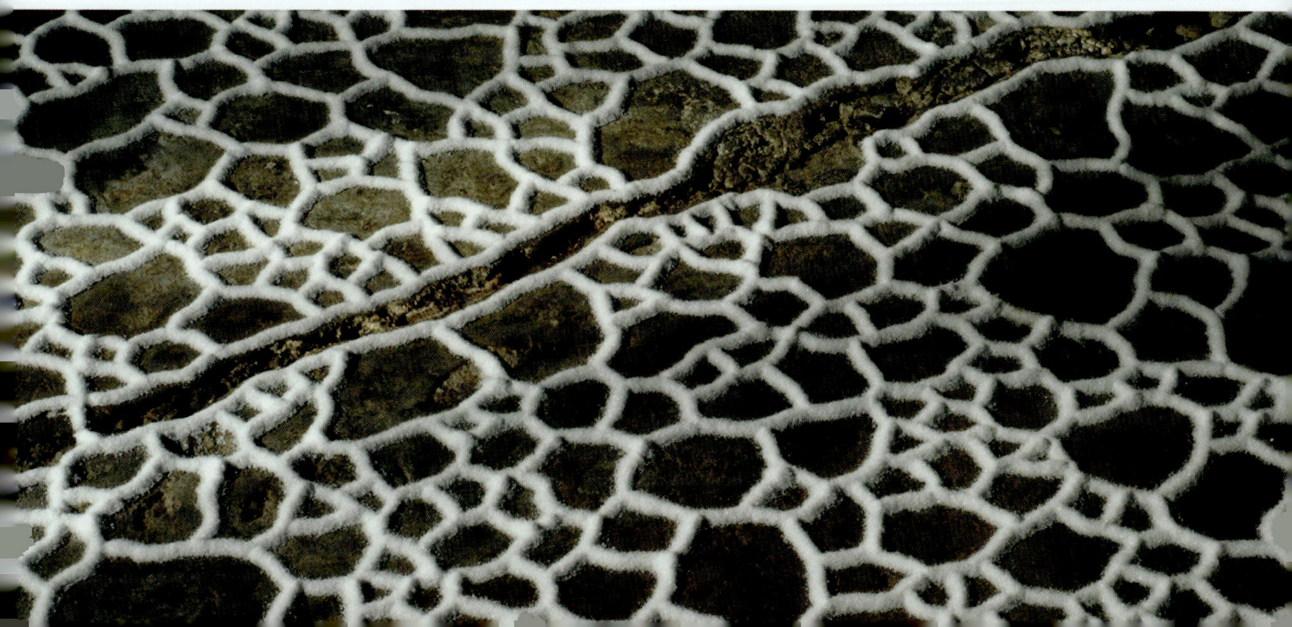

Saltworks

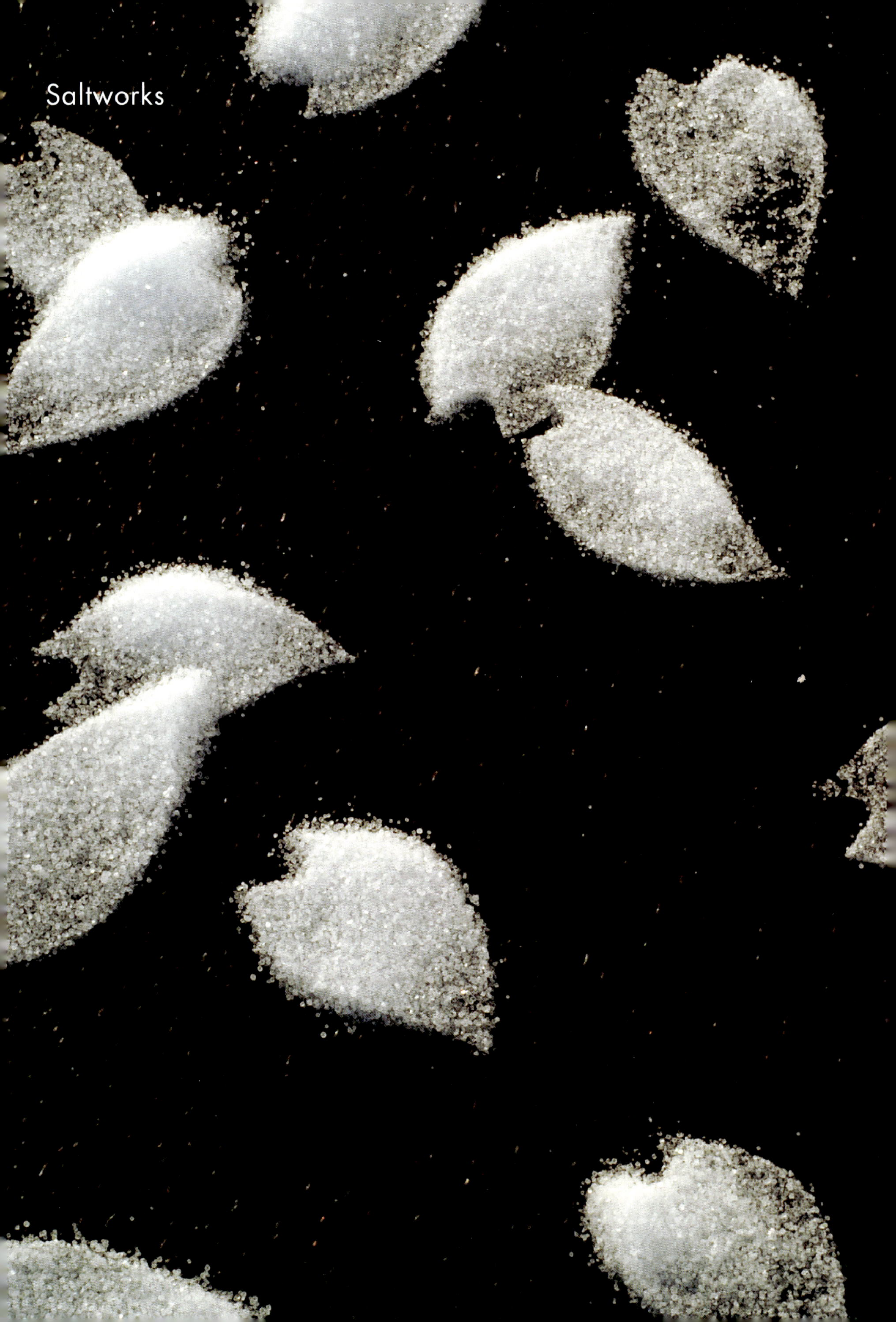

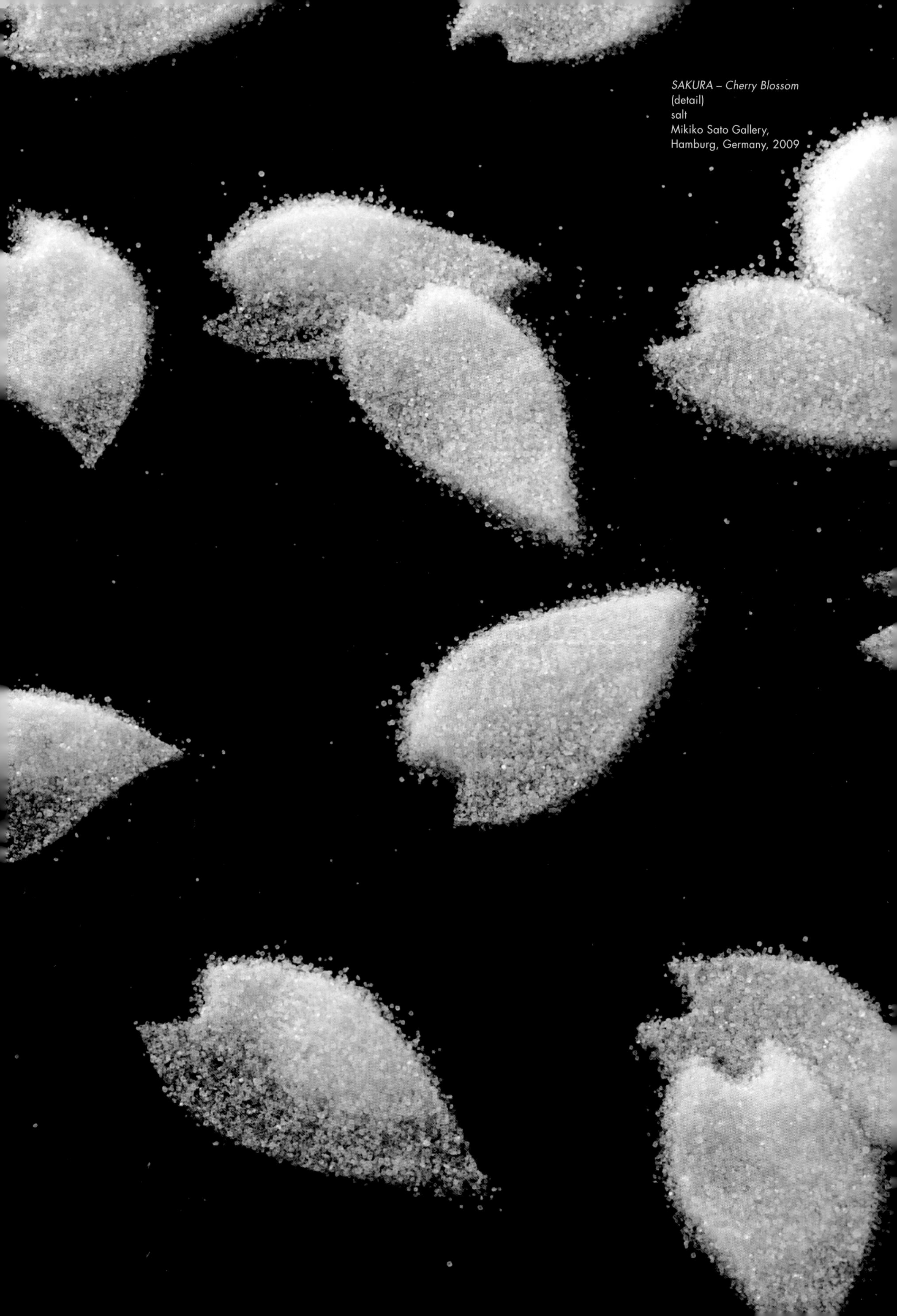

SAKURA – Cherry Blossom
(detail)
salt
Mikiko Sato Gallery,
Hamburg, Germany, 2009

Utsusemi

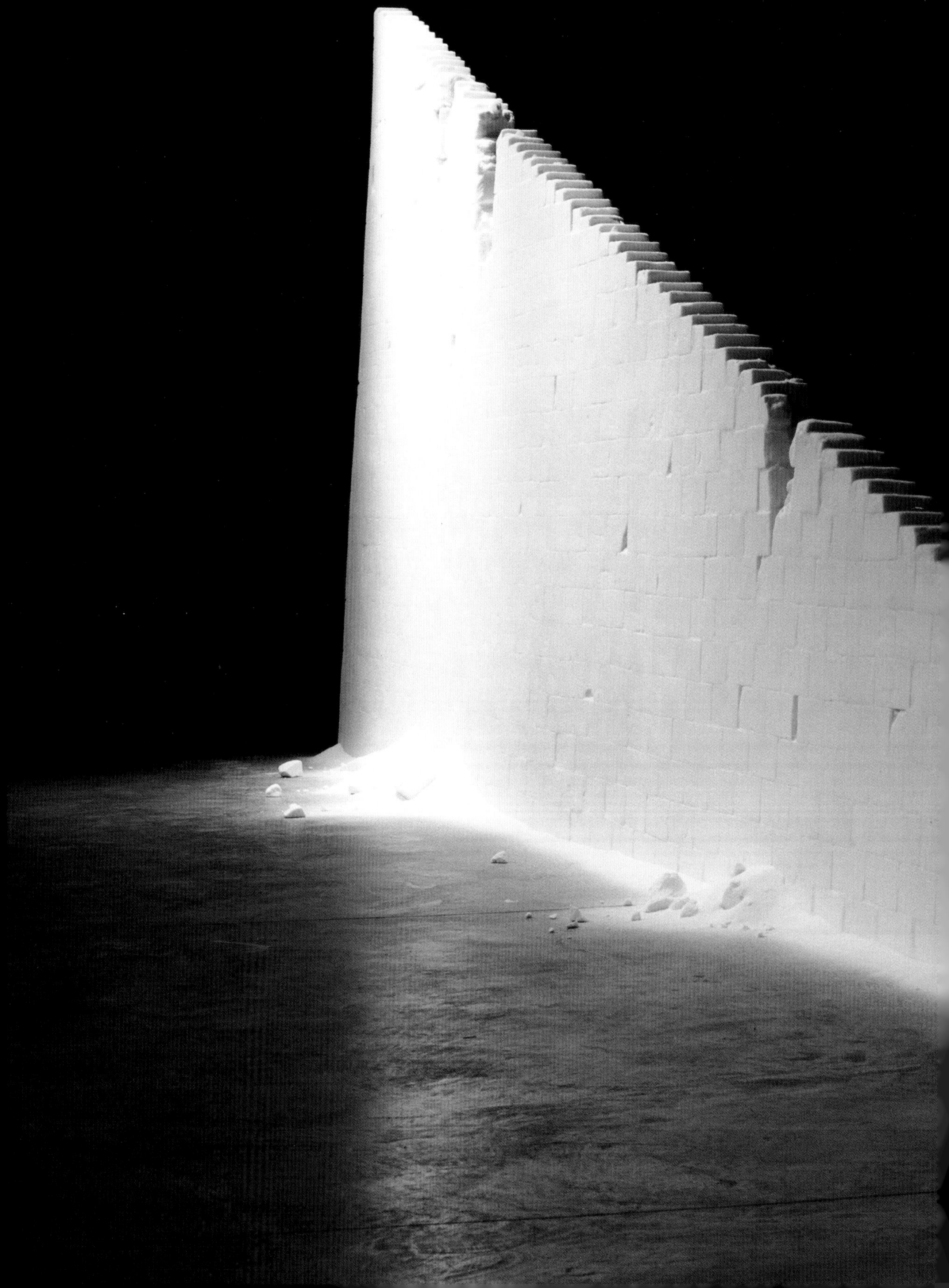

UTSUSEMI
salt / 1.5 x 7.8 x 2.8m
P.S.1 Contemporary Art Center
New York, 2003

UTSUSEMI
salt / 2.0 x 7.5 x 2.5m
Nizayama Forest Art Museum
Toyama, Japan, 2005

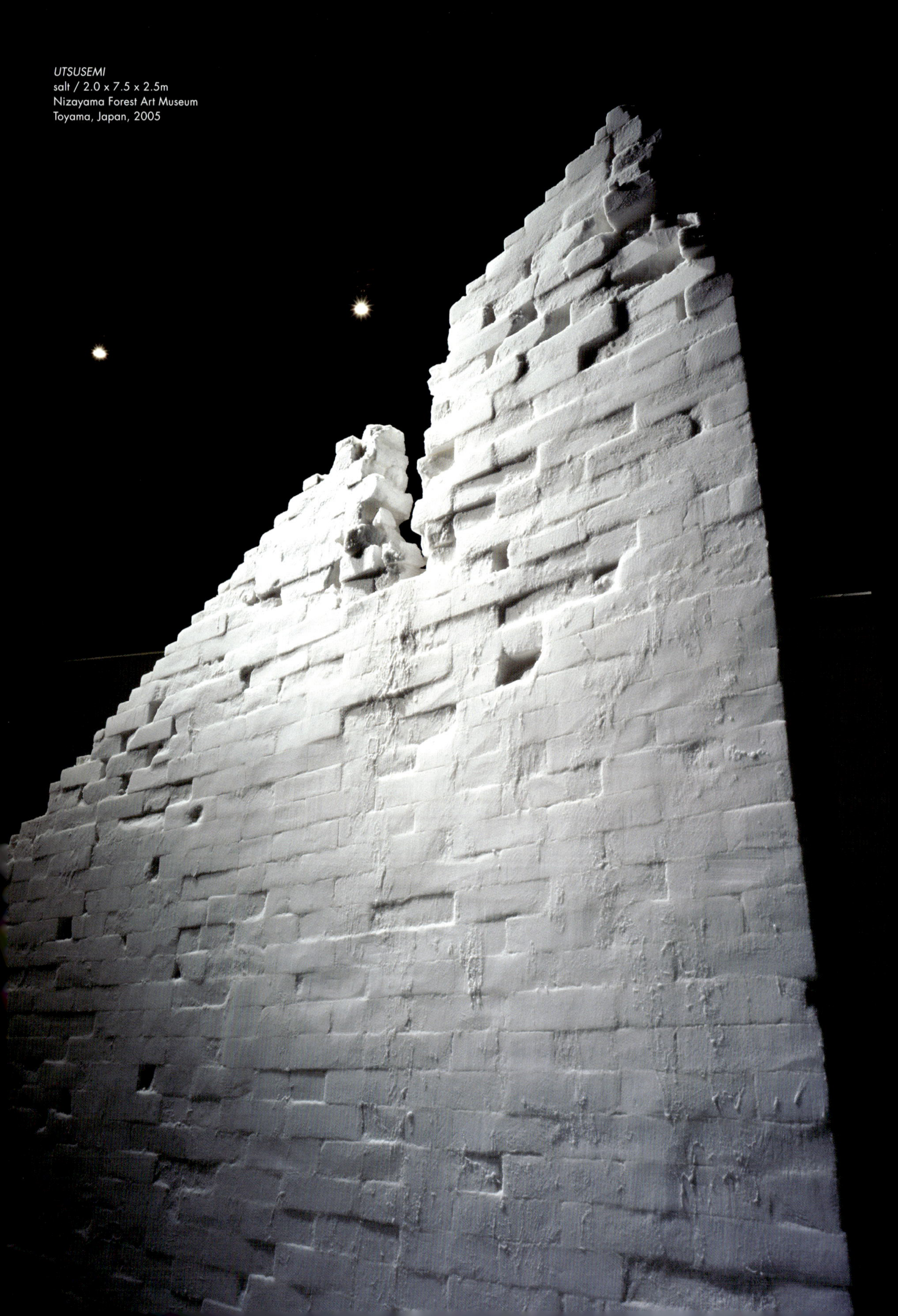

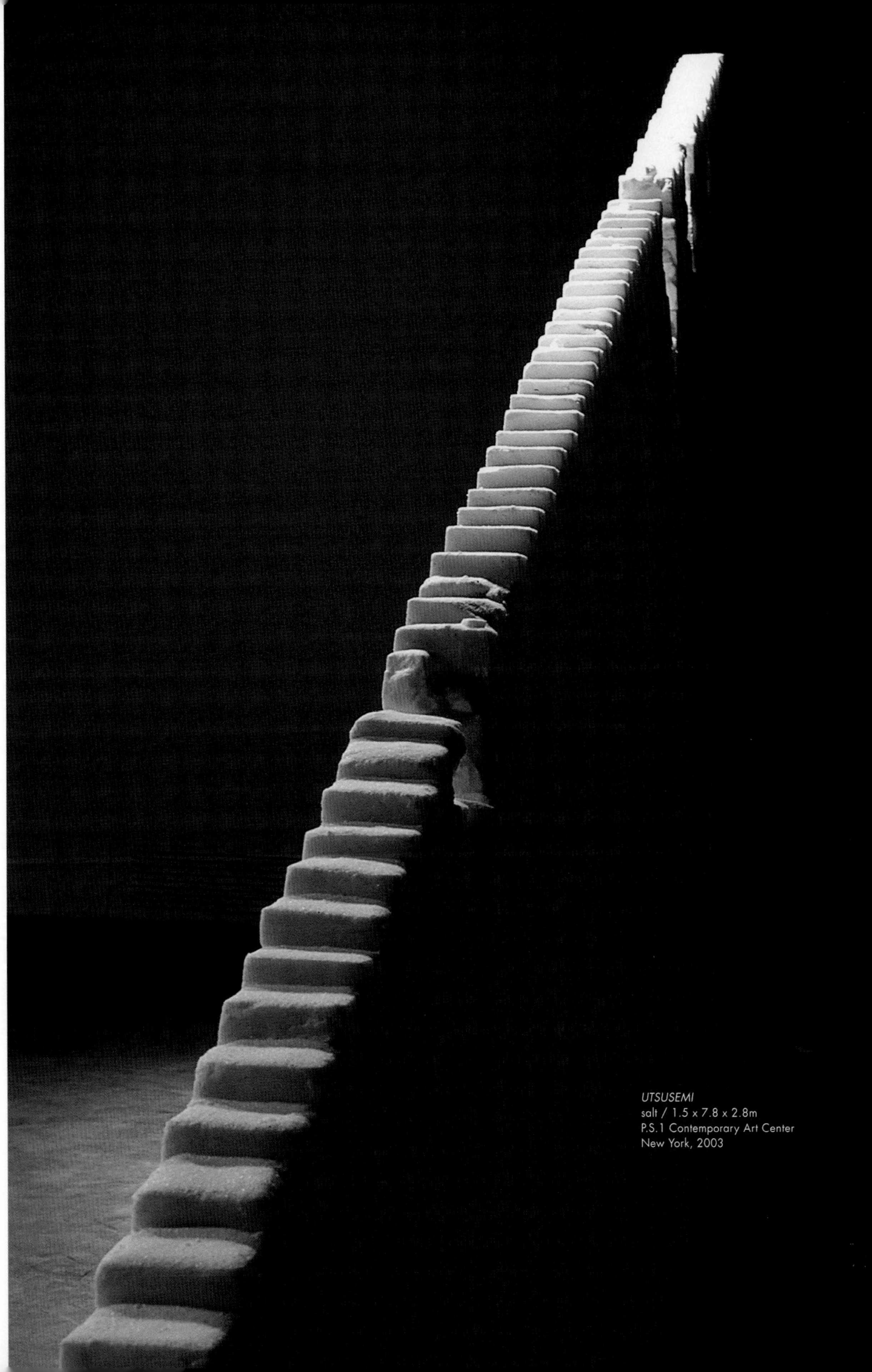

UTSUSEMI
salt / 1.5 x 7.8 x 2.8m
P.S.1 Contemporary Art Center
New York, 2003

A Corridor To Remembrance

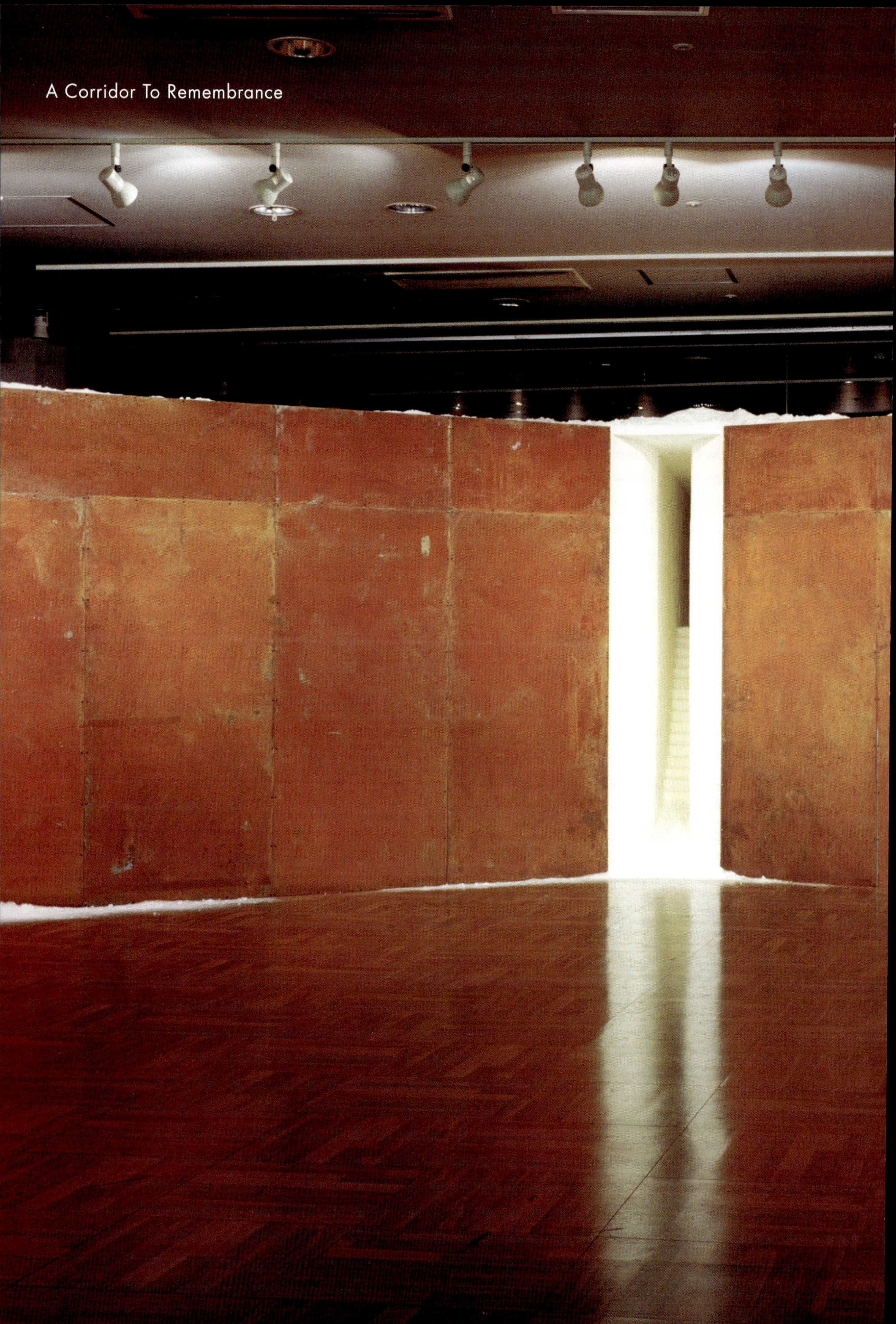

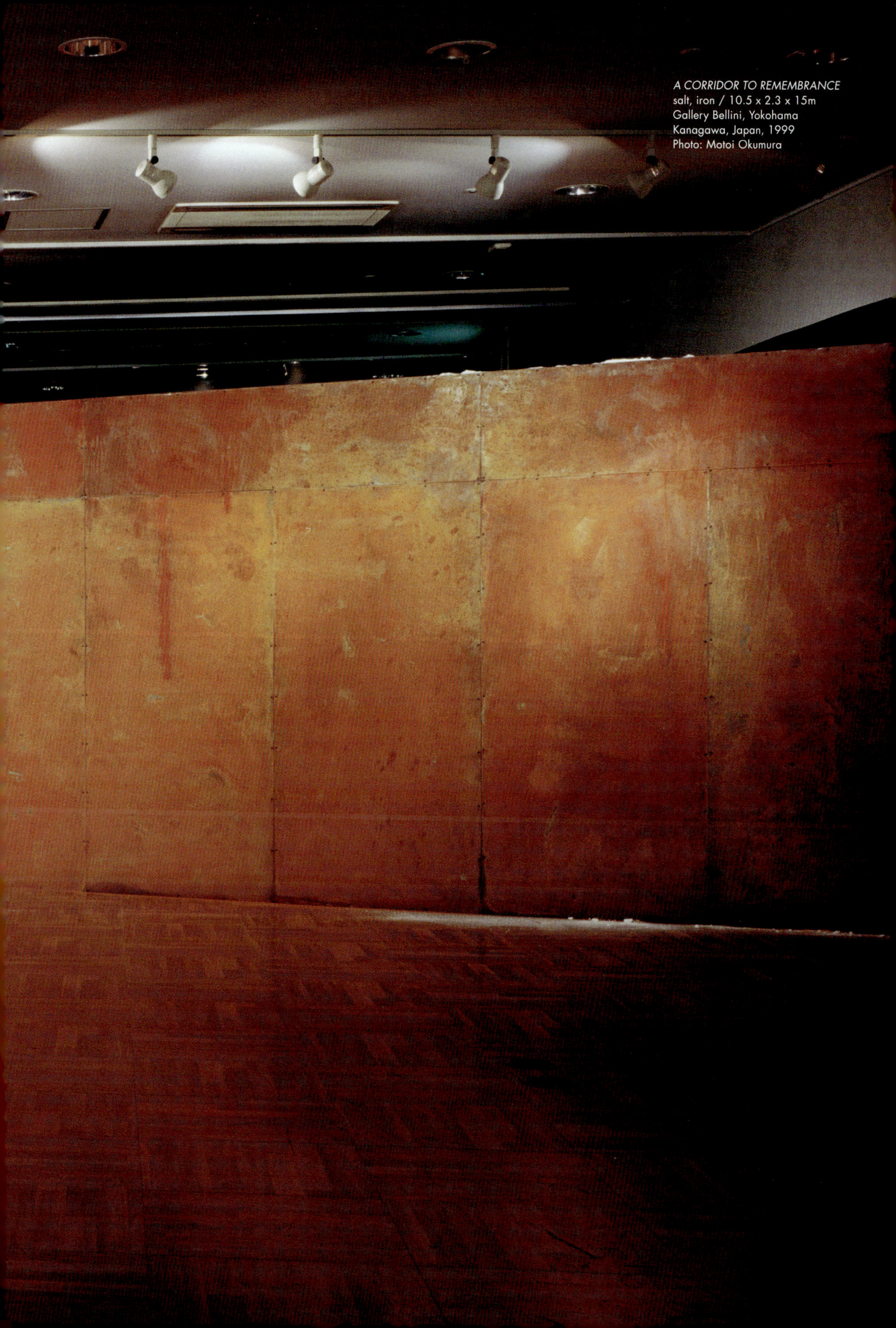

A CORRIDOR TO REMEMBRANCE
salt, iron / 10.5 x 2.3 x 15m
Gallery Bellini, Yokohama
Kanagawa, Japan, 1999
Photo: Motoi Okumura

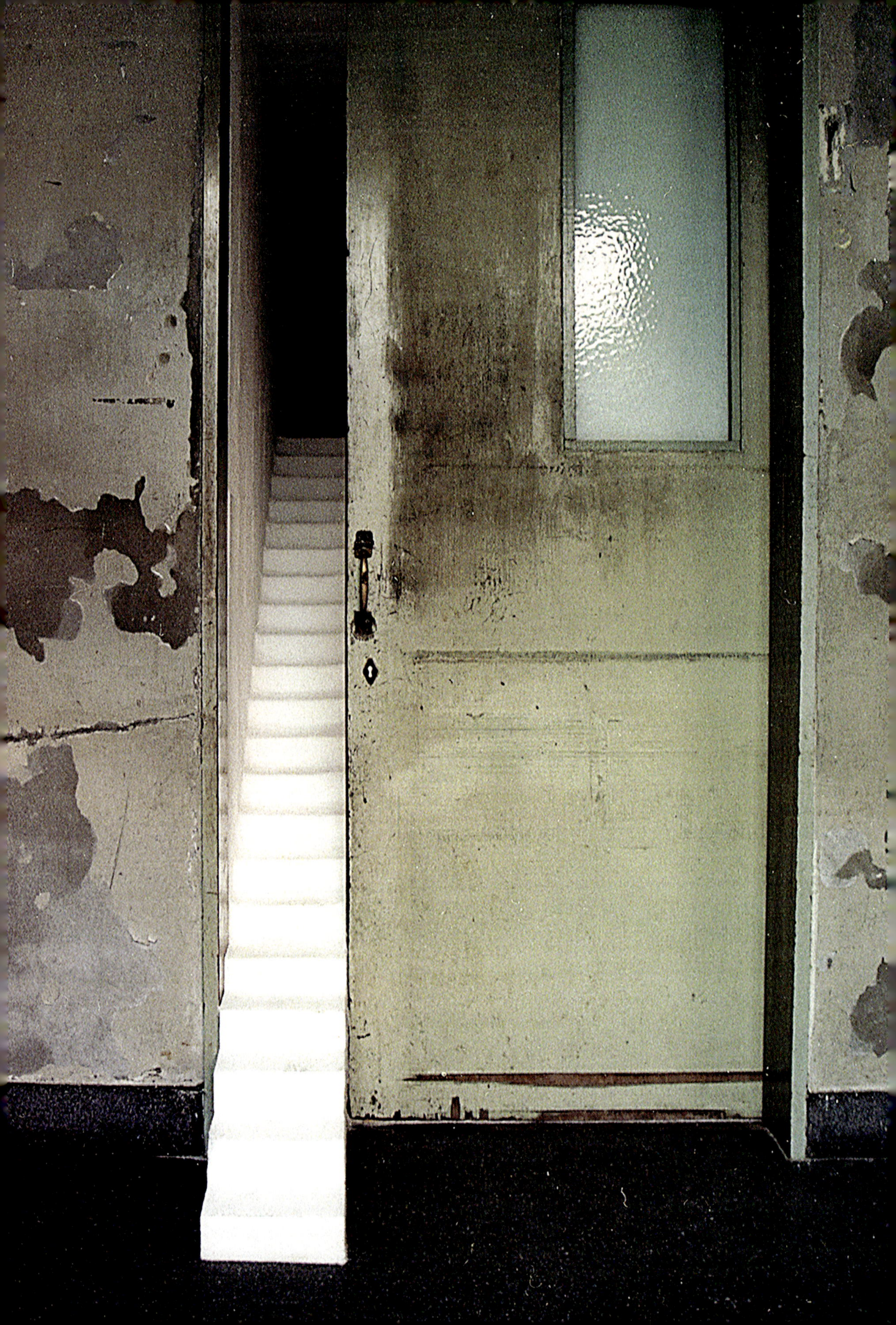

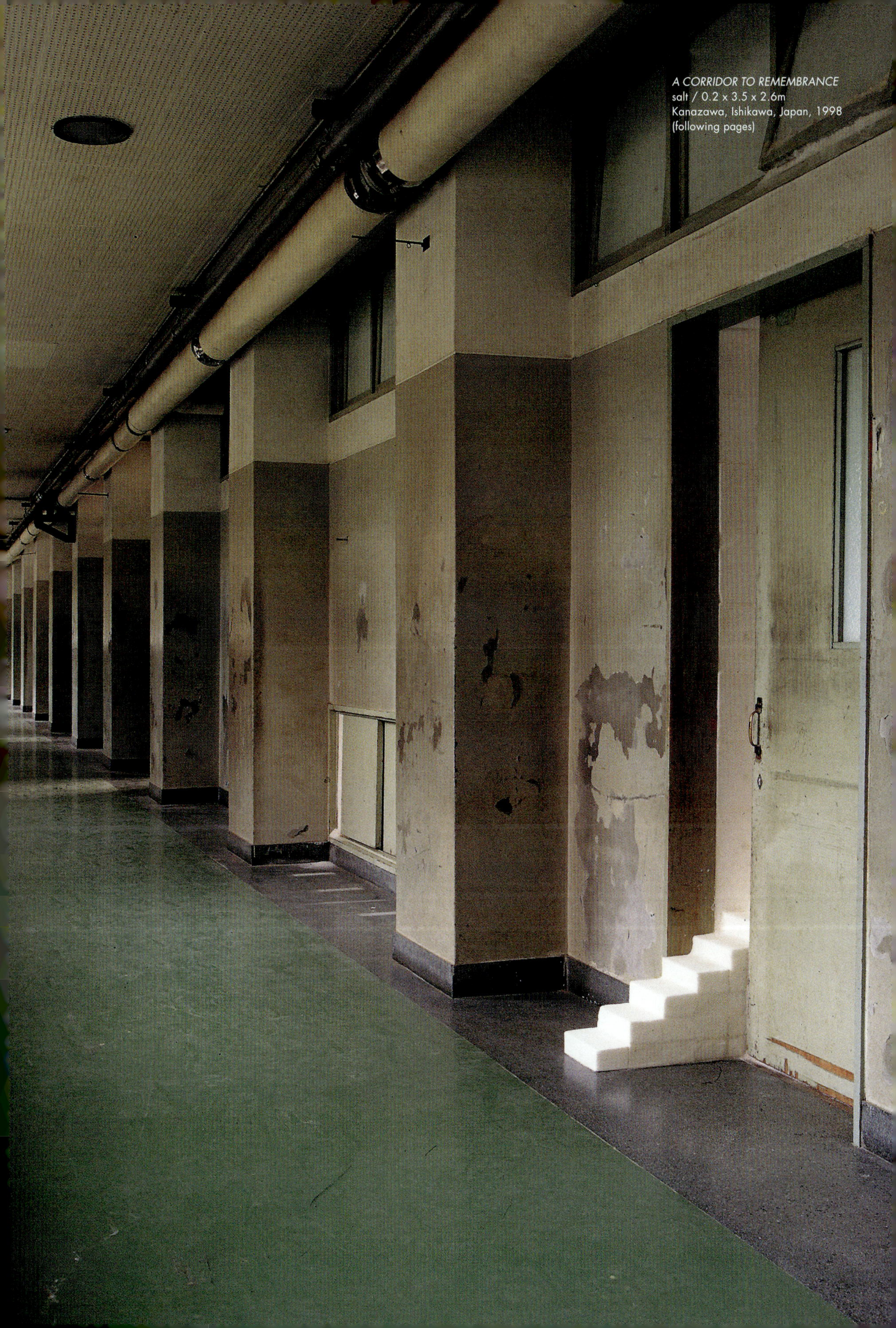

A CORRIDOR TO REMEMBRANCE
salt / 0.2 x 3.5 x 2.6m
Kanazawa, Ishikawa, Japan, 1998
(following pages)

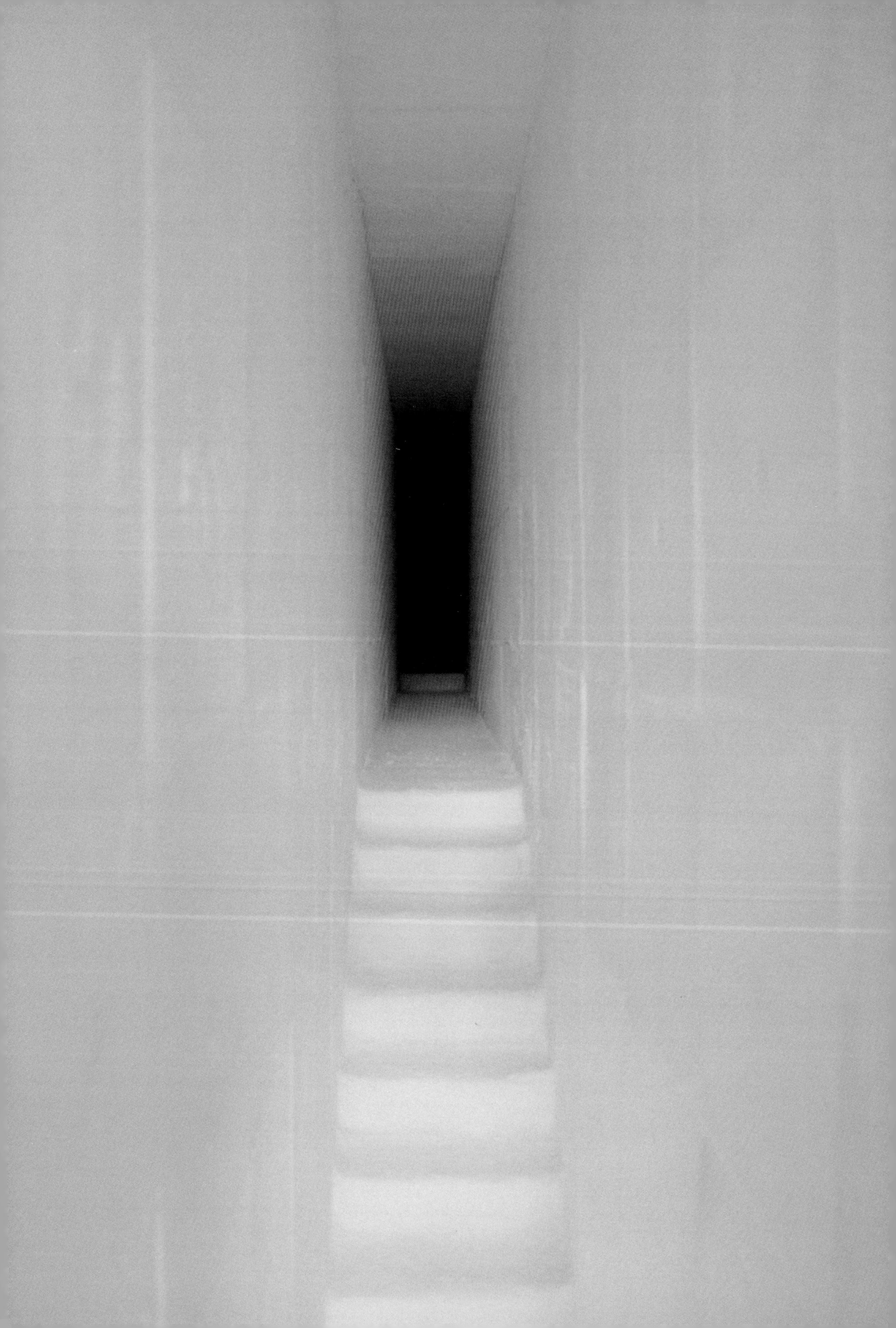

A CORRIDOR TO REMEMBRANCE
salt, iron / 8 x 3 x 2.3m
Gallery Le Deco
Tokyo, Japan, 1999

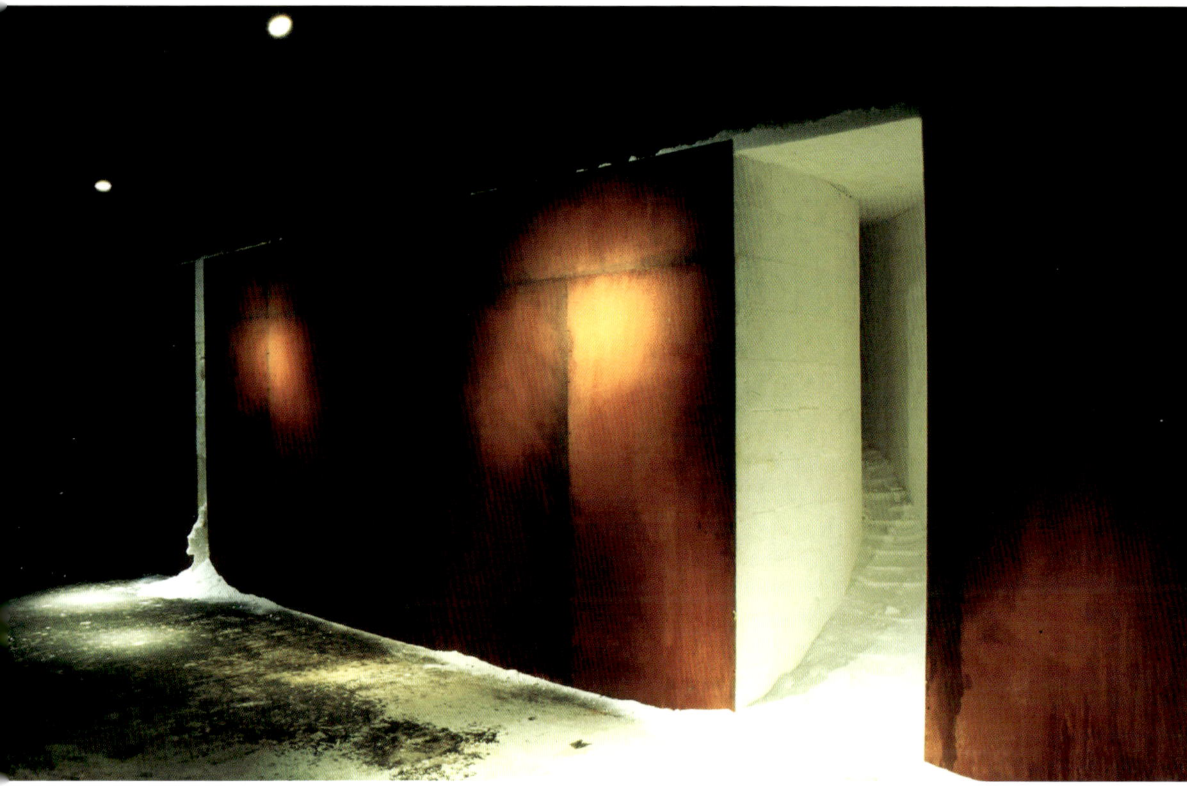

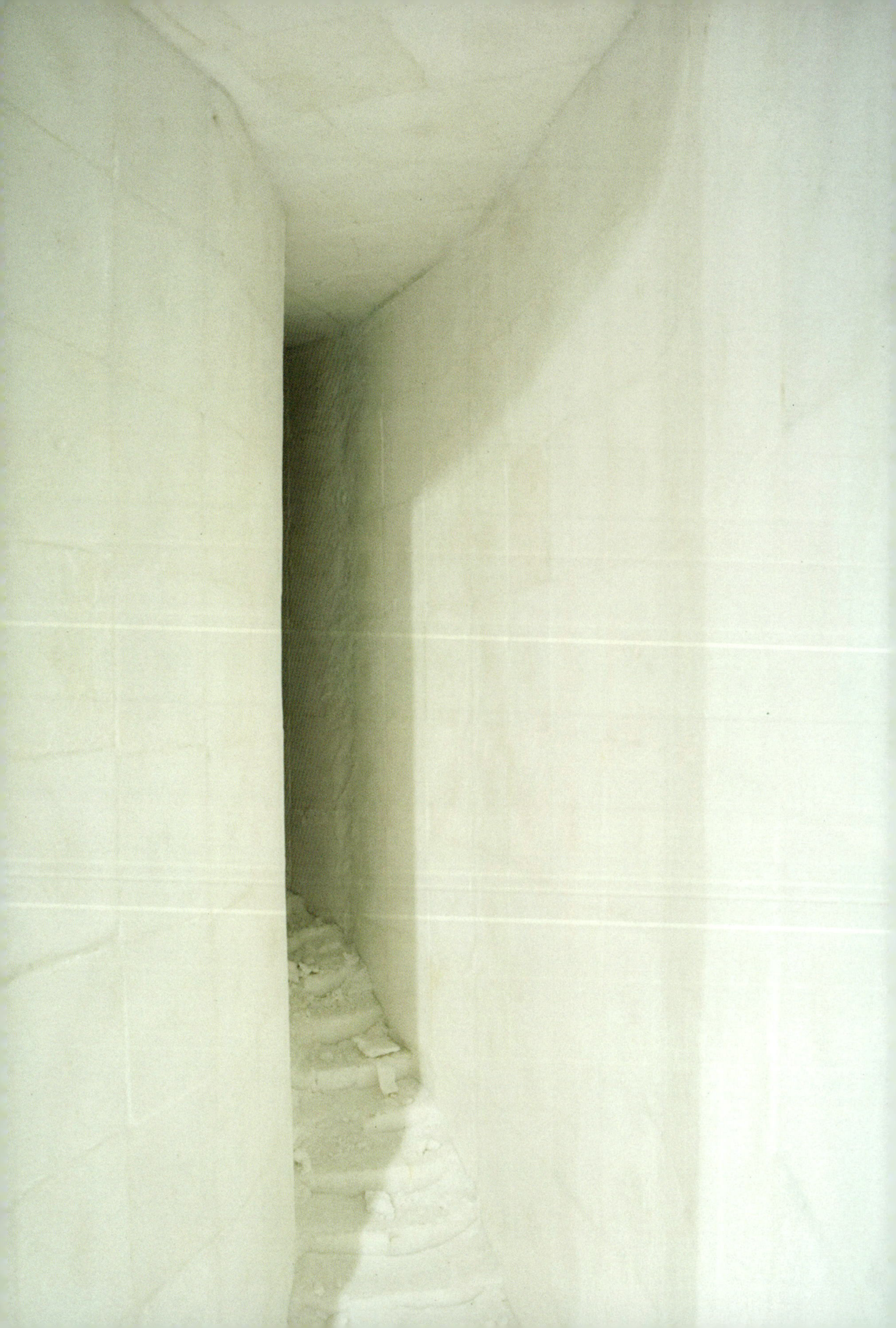

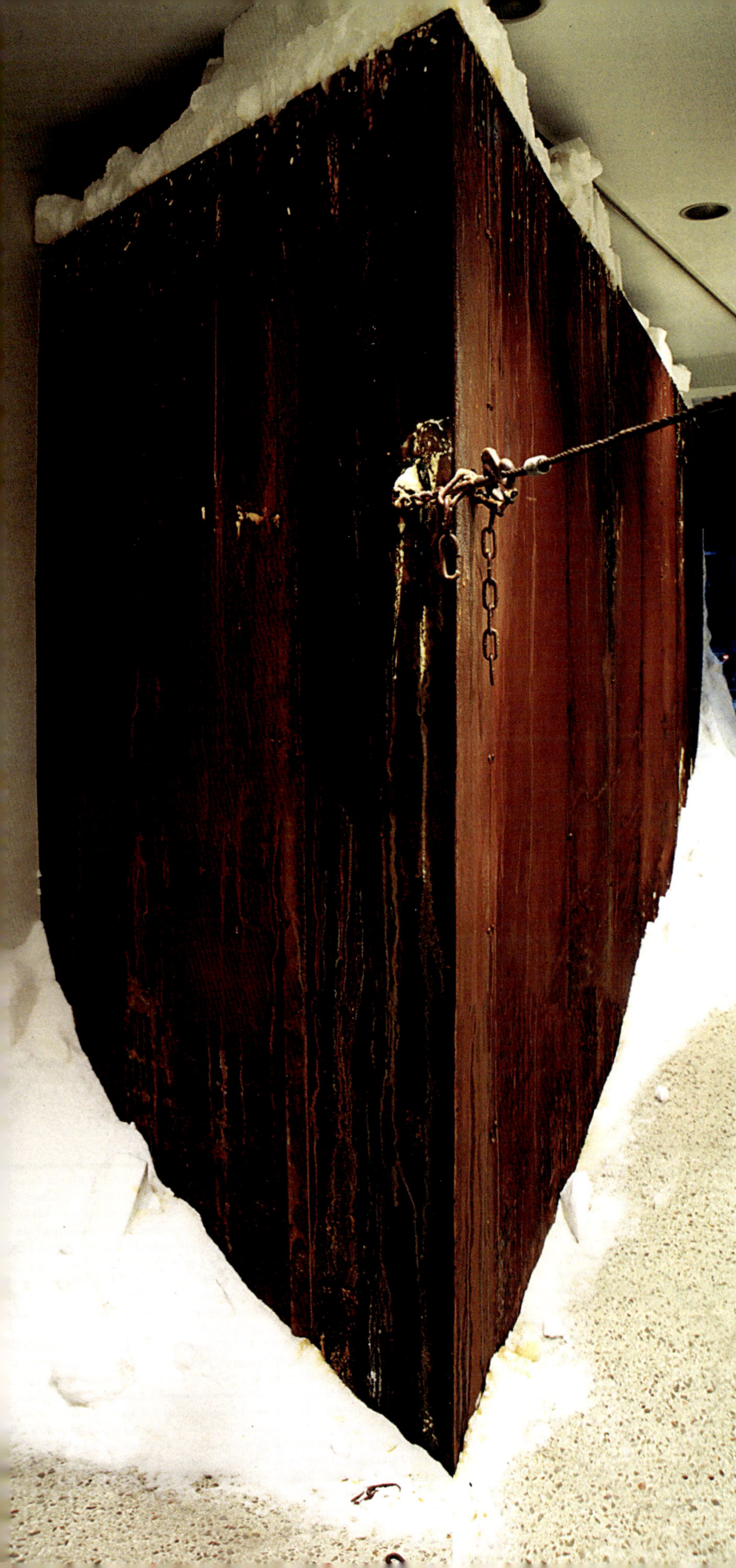

UNFULFILLED
salt, iron
Gallery G2
Fukui, Japan, 2000
(and following pages)

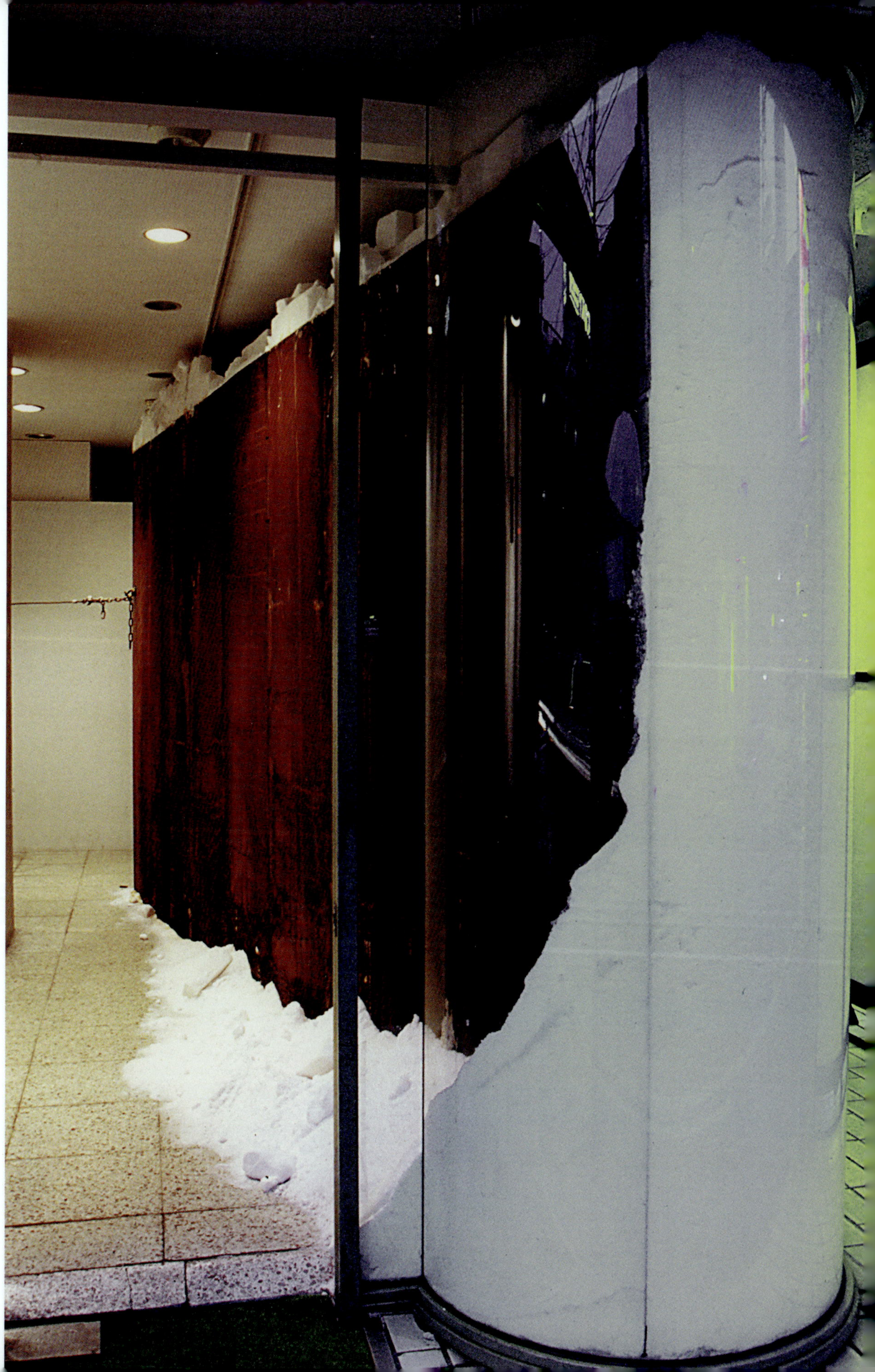

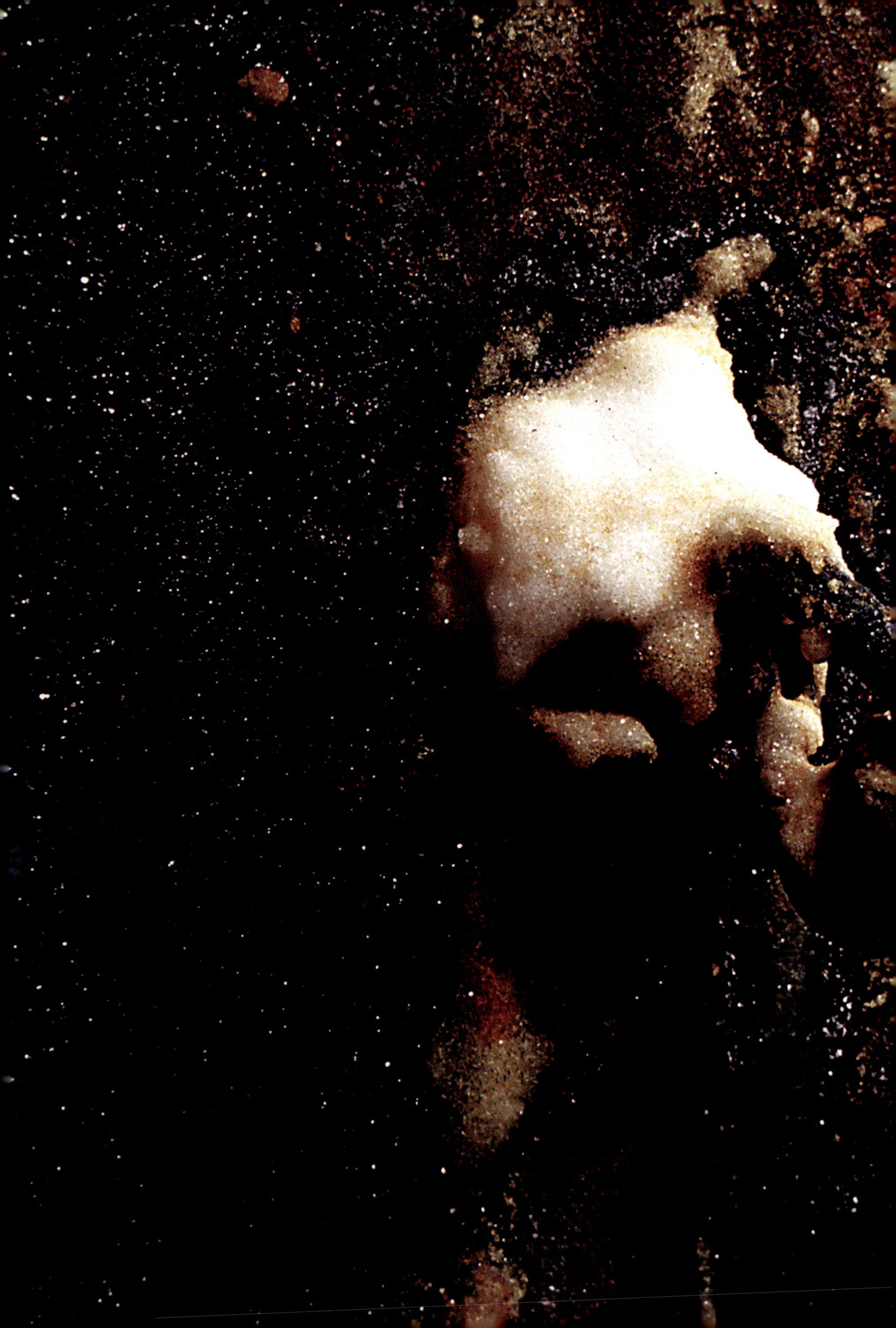

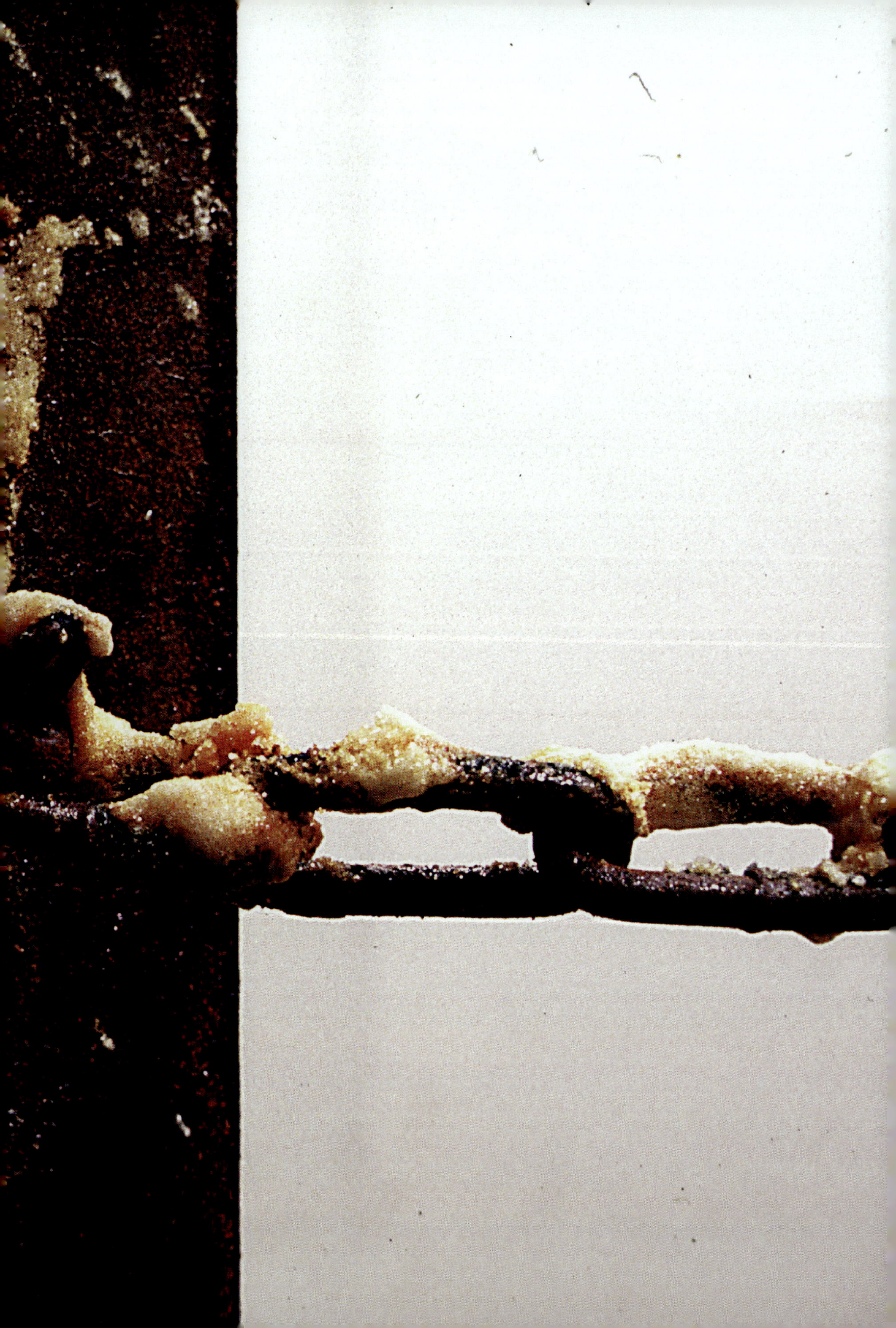

Sakura

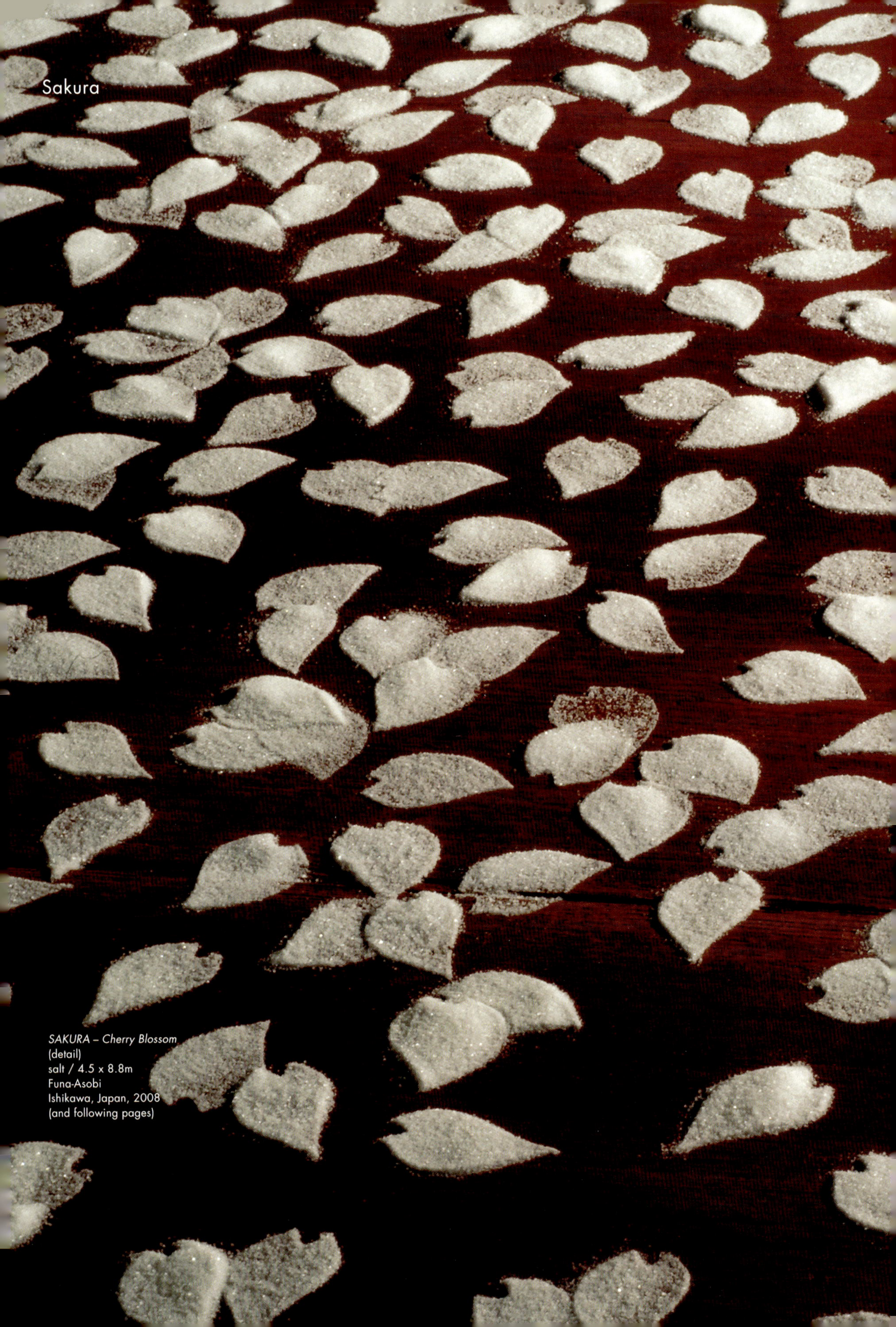

SAKURA – Cherry Blossom
(detail)
salt / 4.5 x 8.8m
Funa-Asobi
Ishikawa, Japan, 2008
(and following pages)

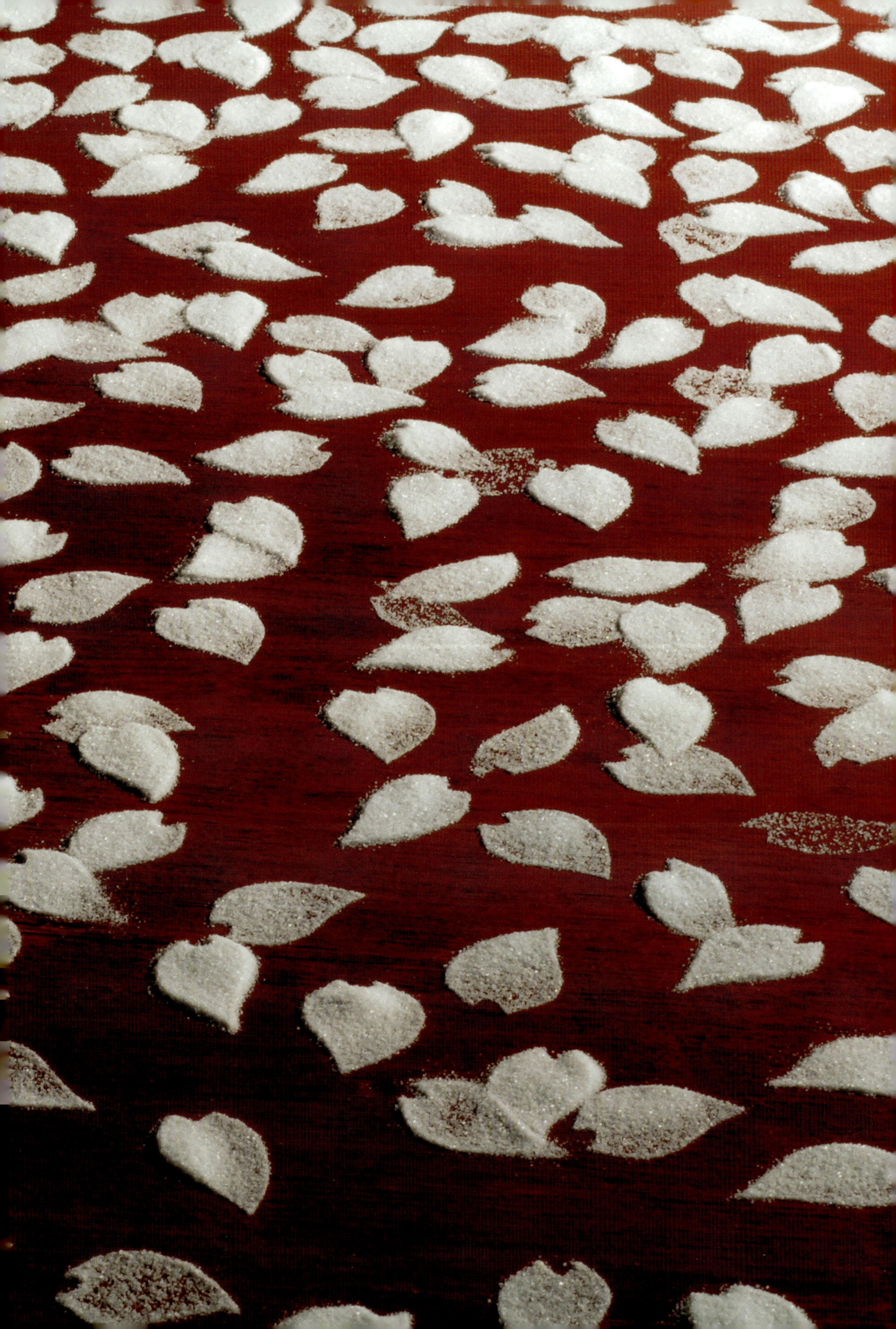

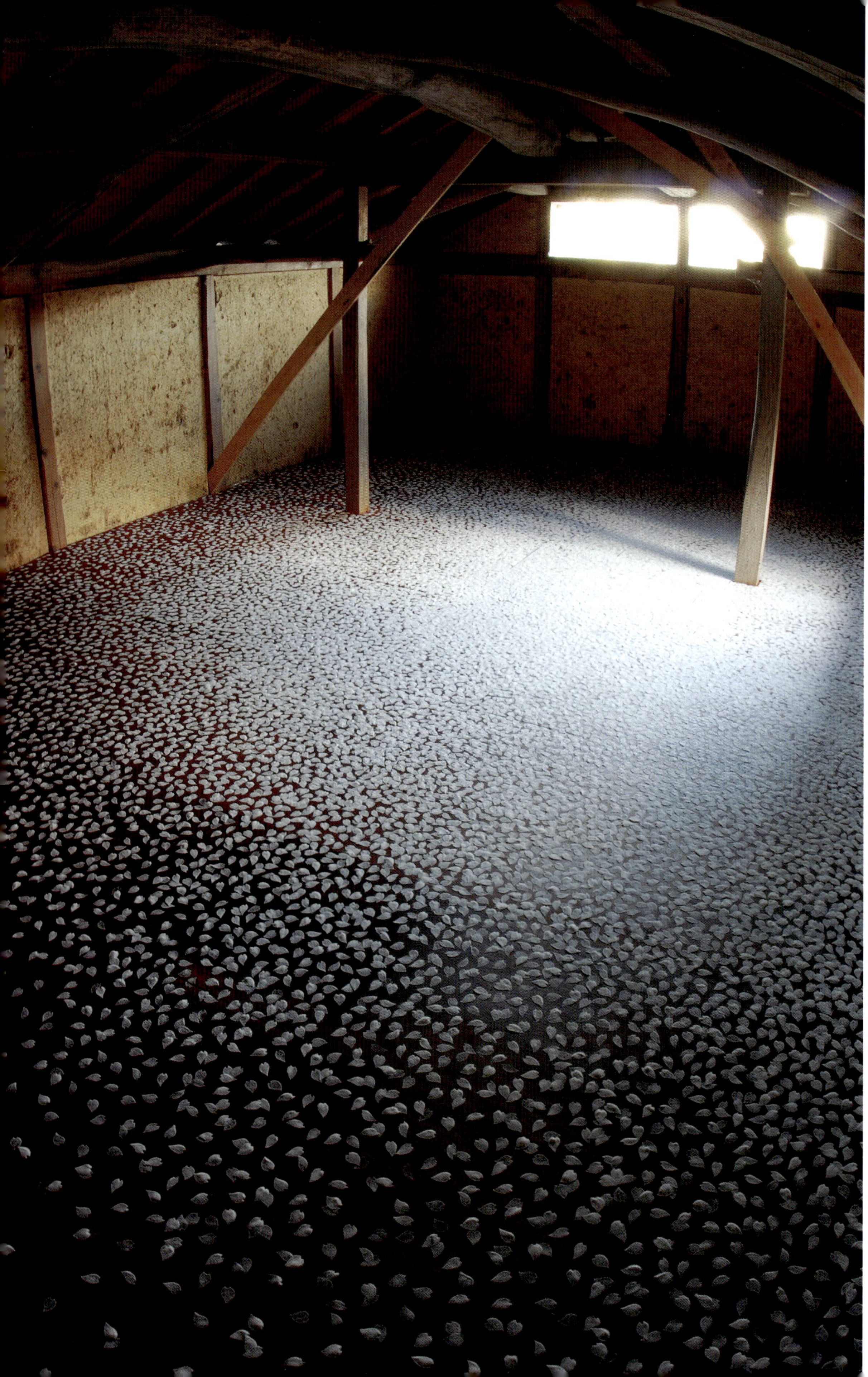

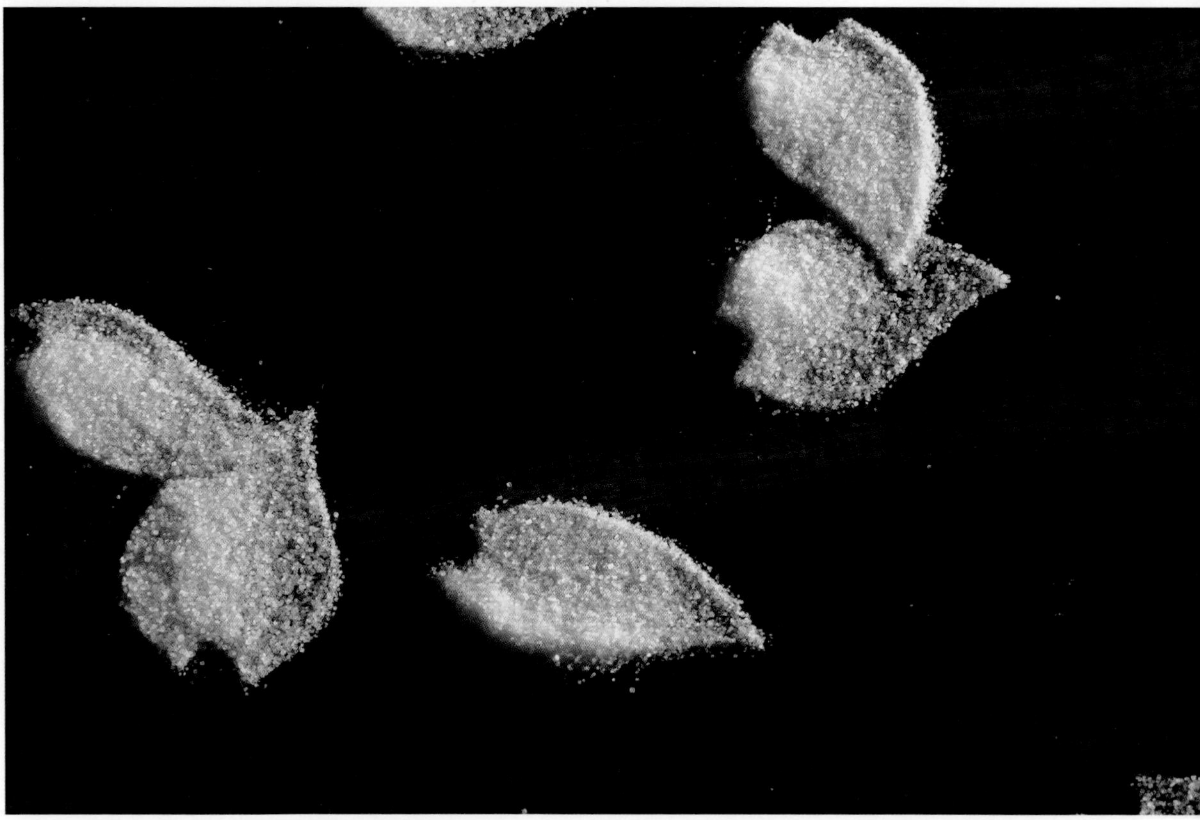

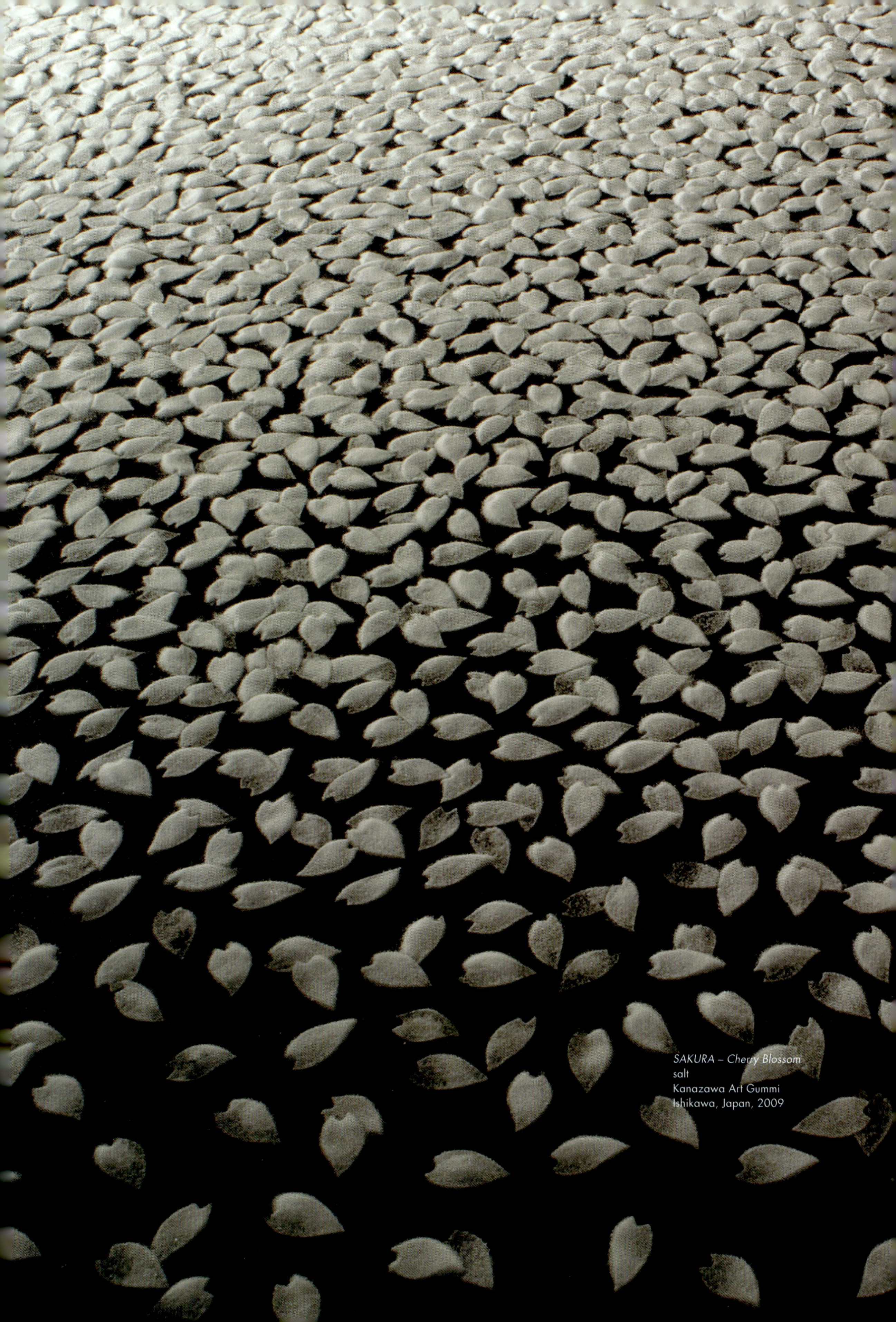

SAKURA – Cherry Blossom
salt
Kanazawa Art Gummi
Ishikawa, Japan, 2009

Labyrinth

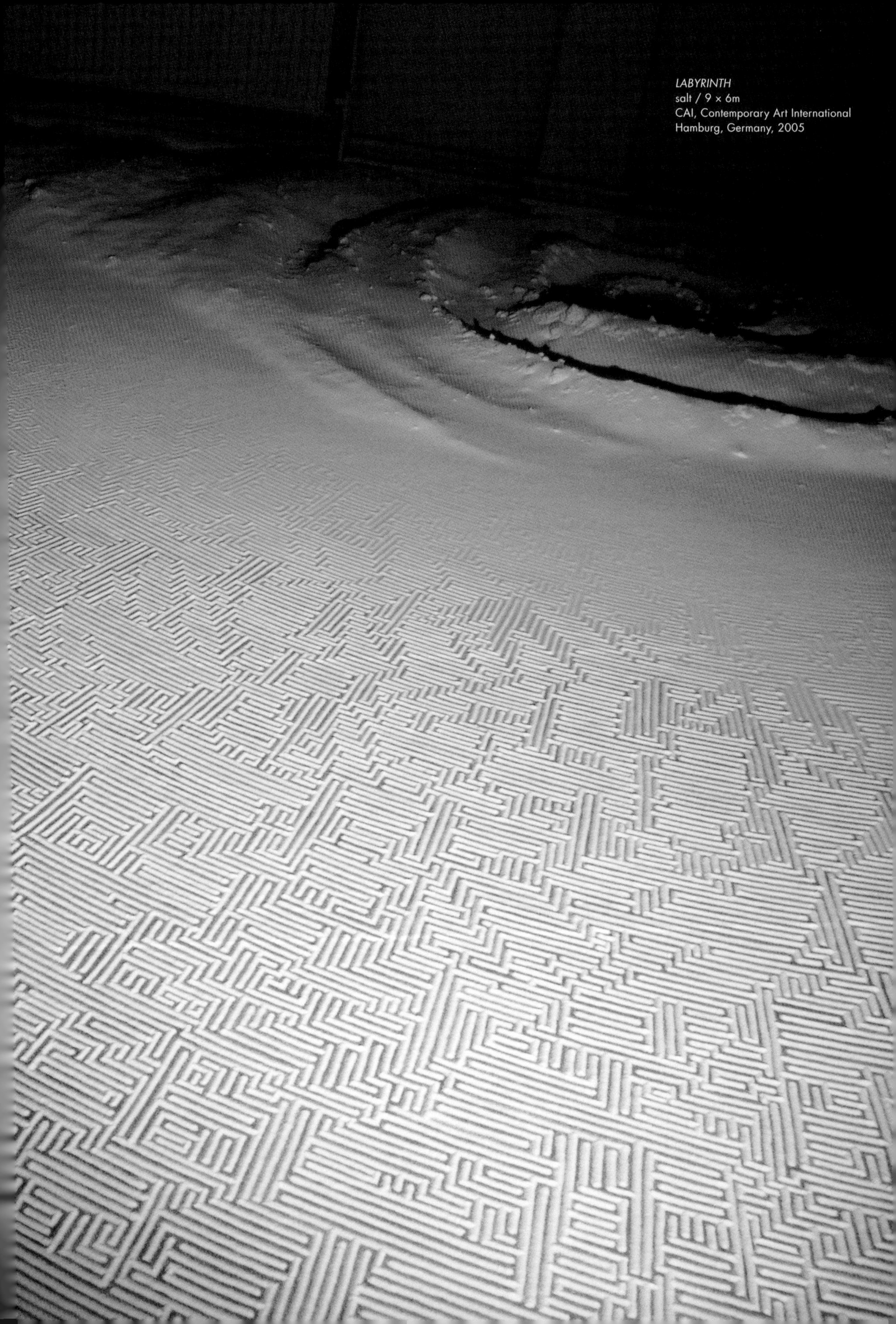

LABYRINTH
salt / 9 × 6m
CAI, Contemporary Art International
Hamburg, Germany, 2005

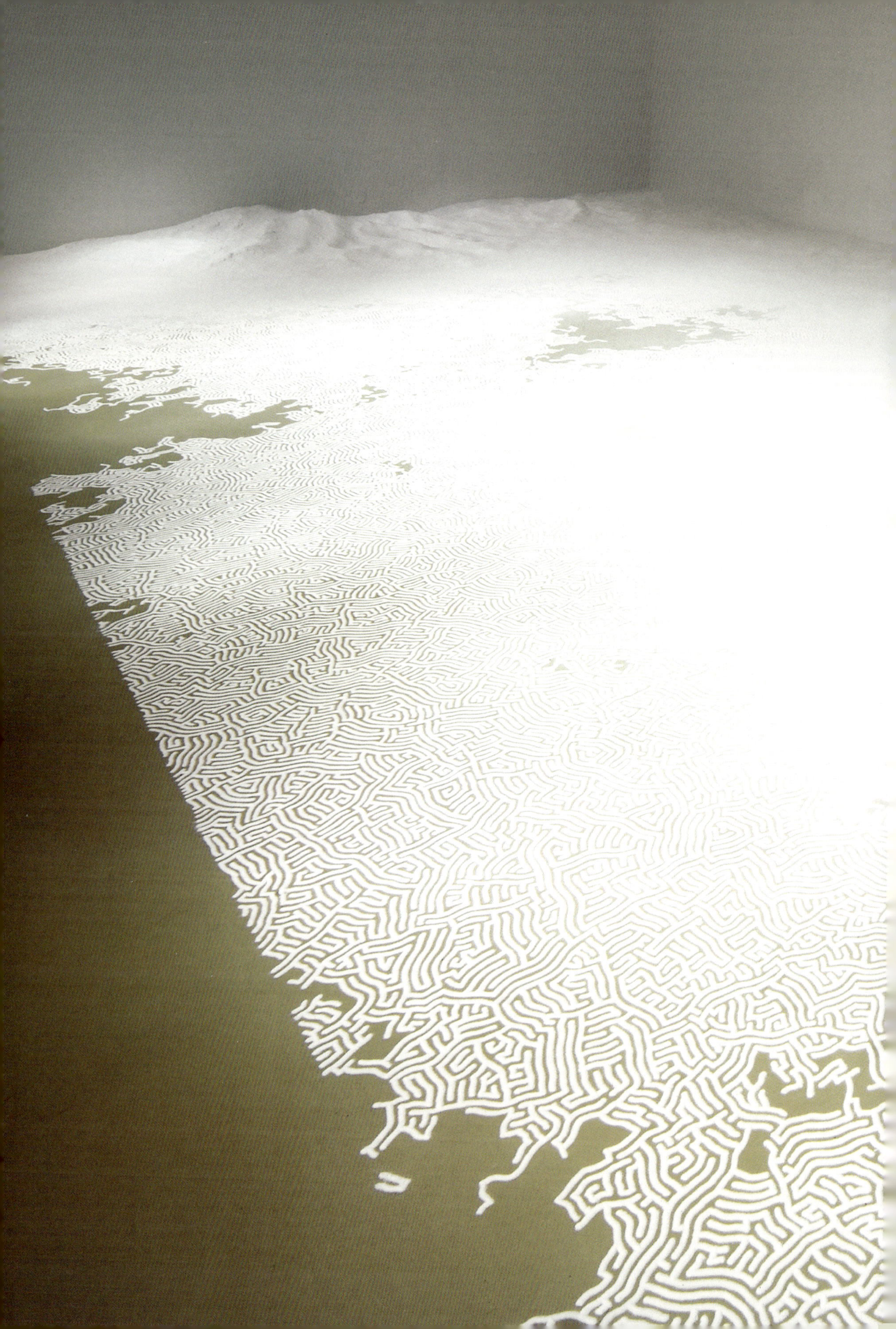

LABYRINTH
salt / 5.5 x 11m
The Israel Museum, Jerusalem
Jerusalem, Israel, 2005
Photo: The Israel Museum, Jerusalem
(and following pages)

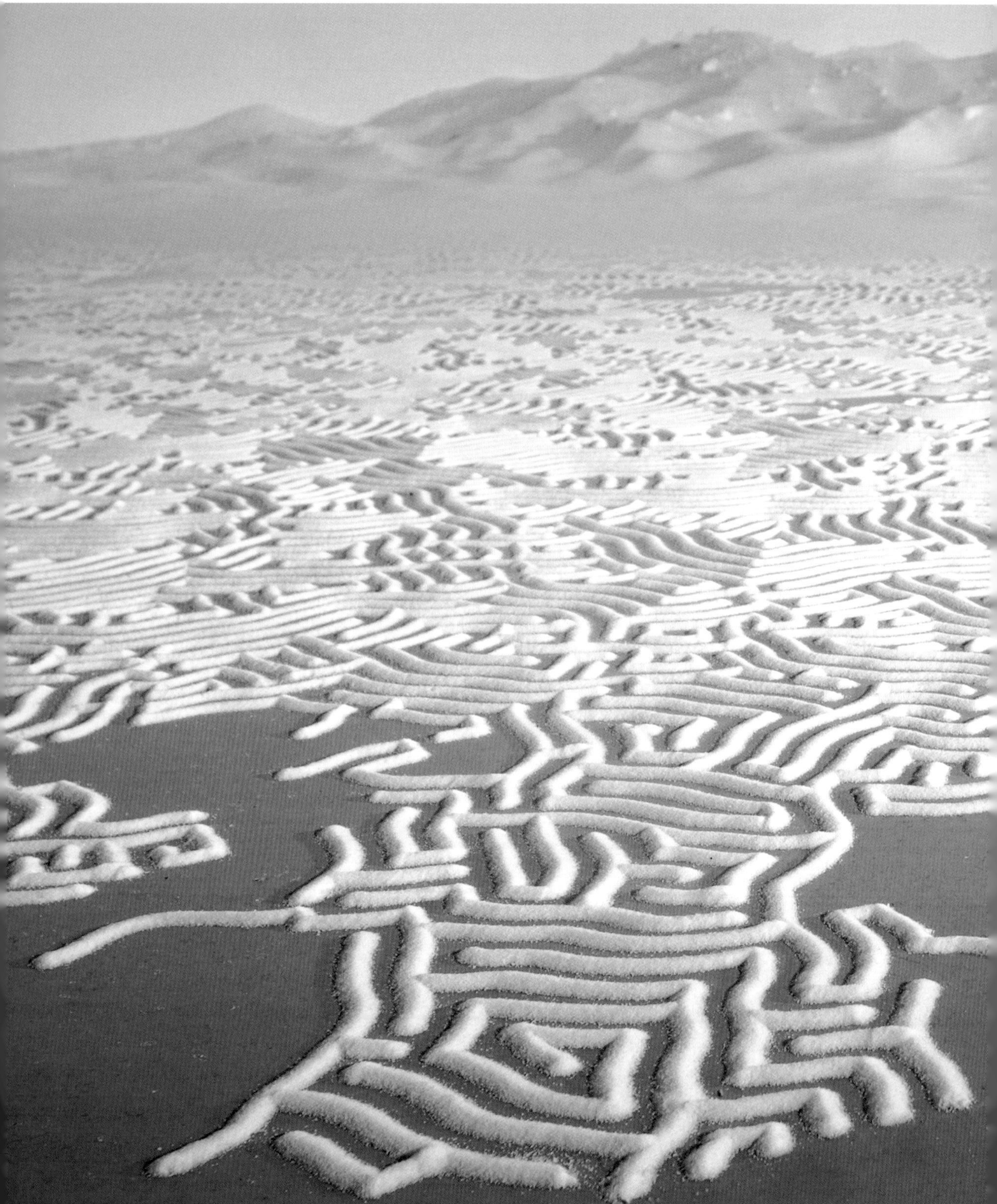

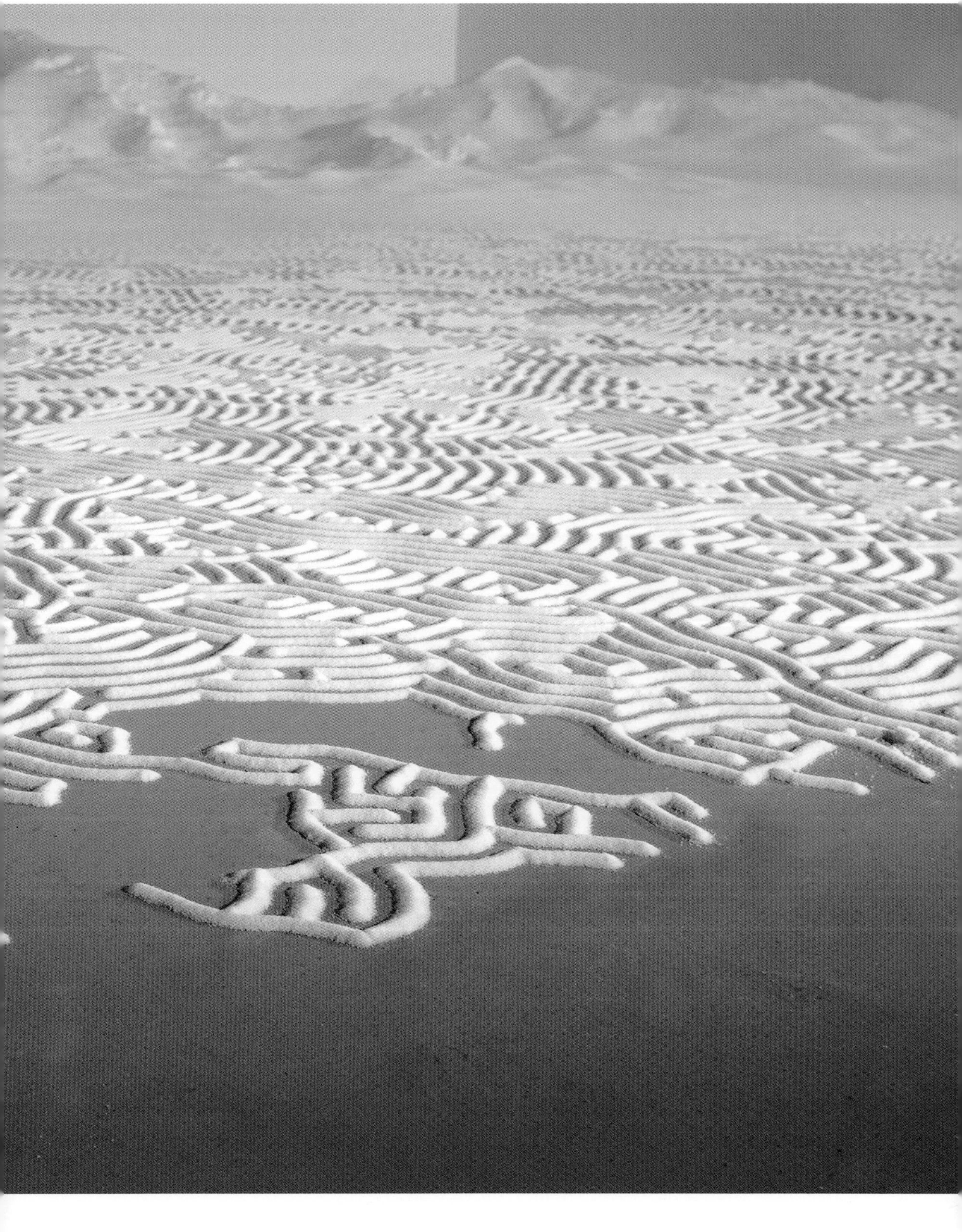

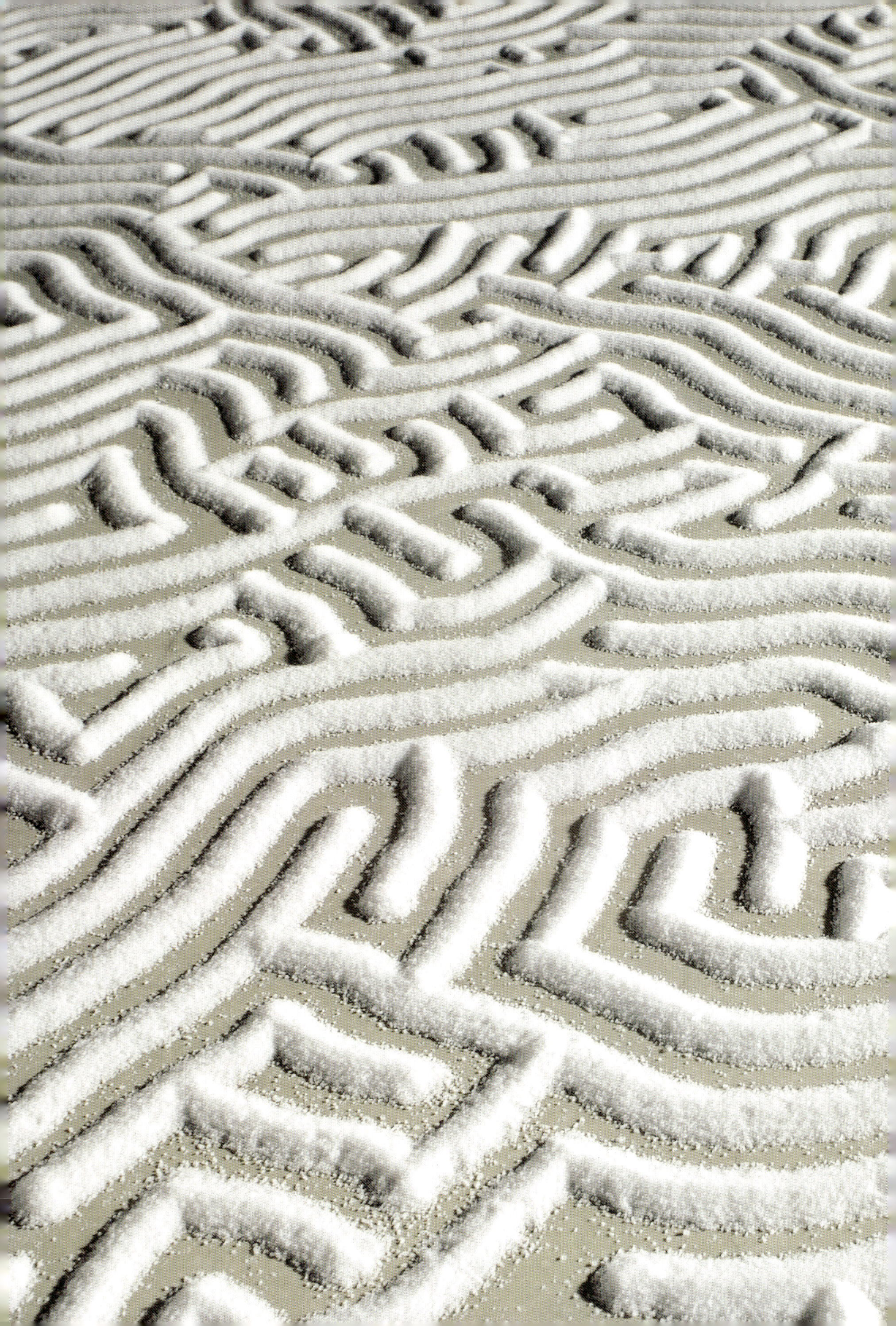

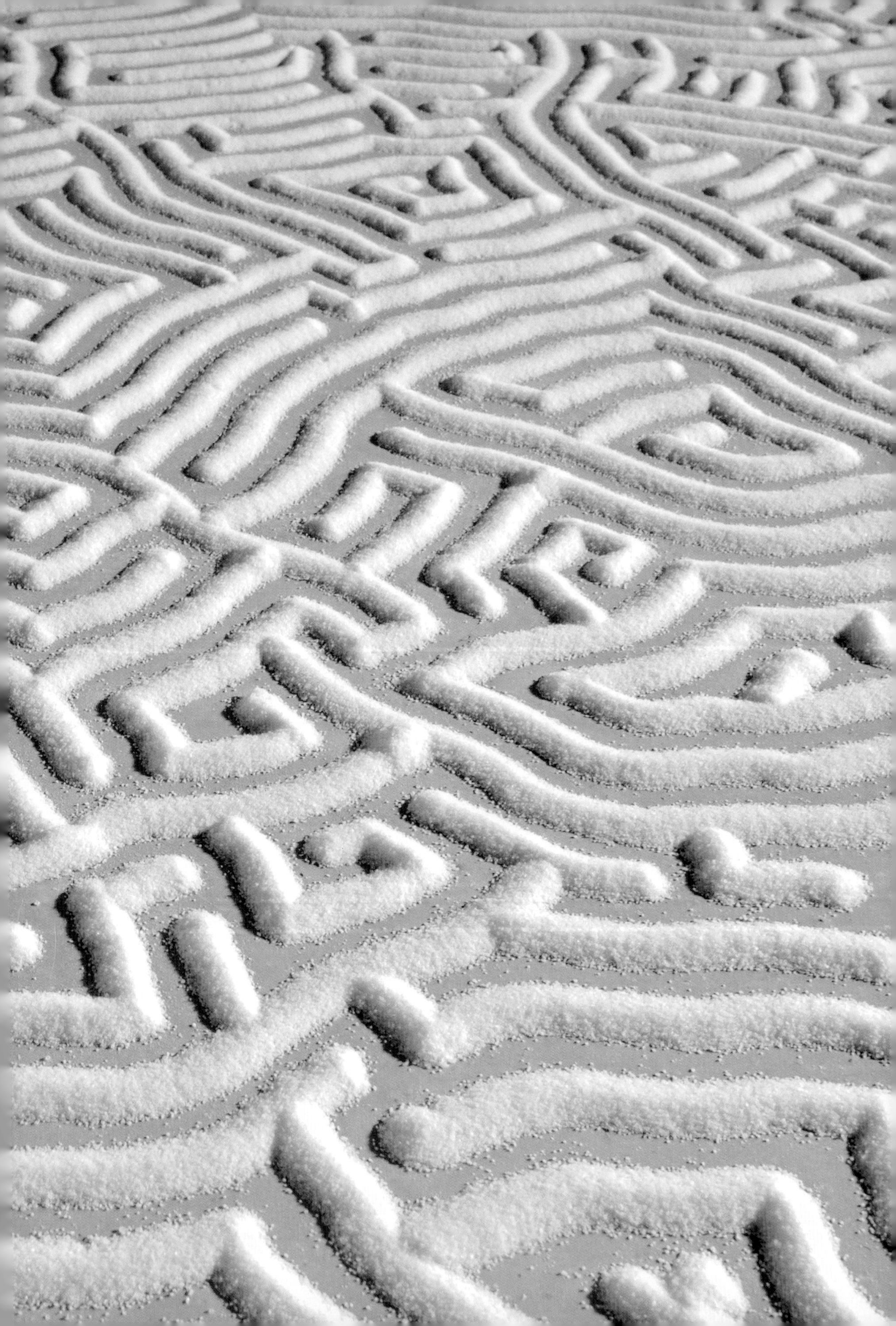

LABYRINTH
salt / 4 × 8.5m
Gallery Sowaka
Kyoto, Japan, 2005
(and following pages)

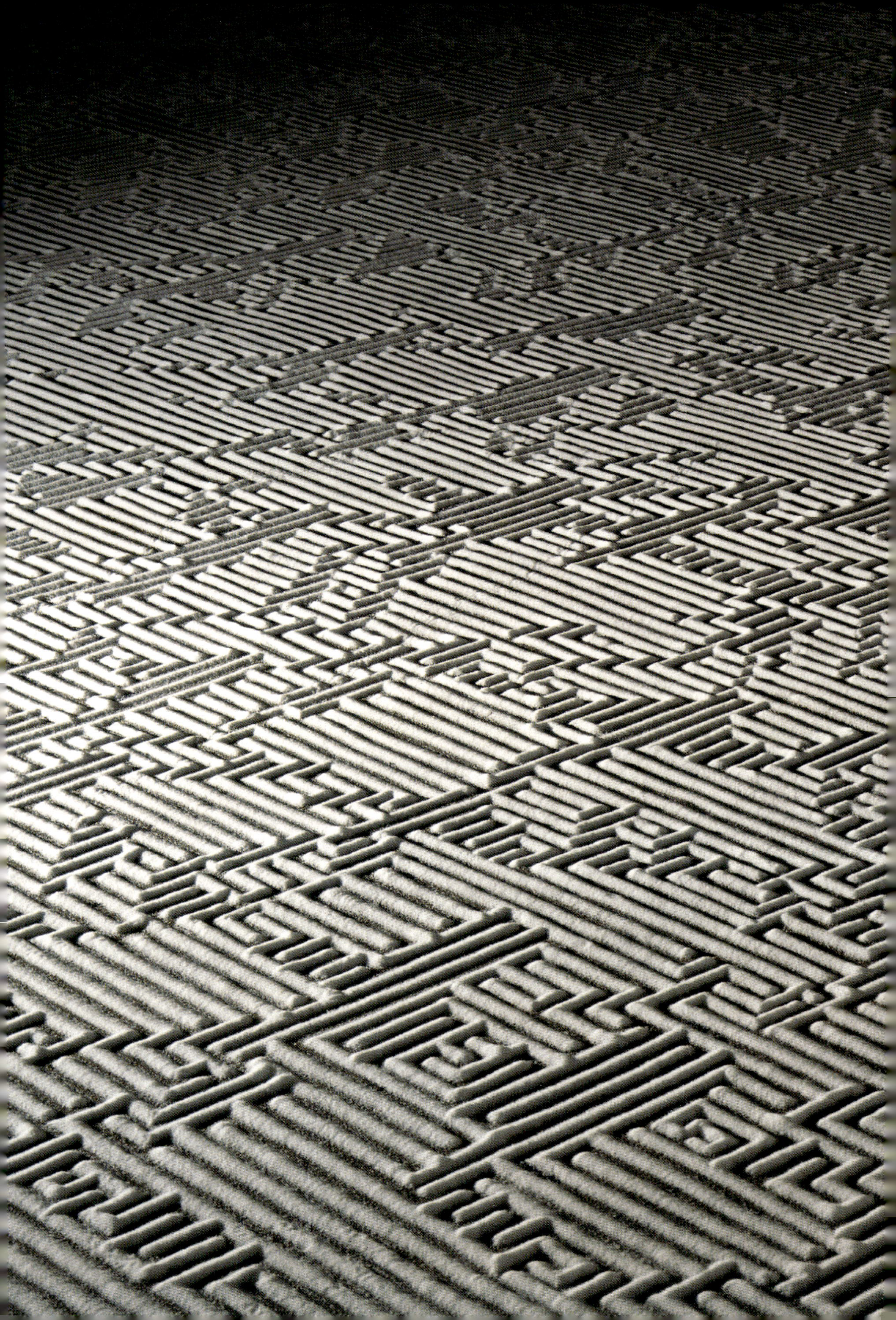

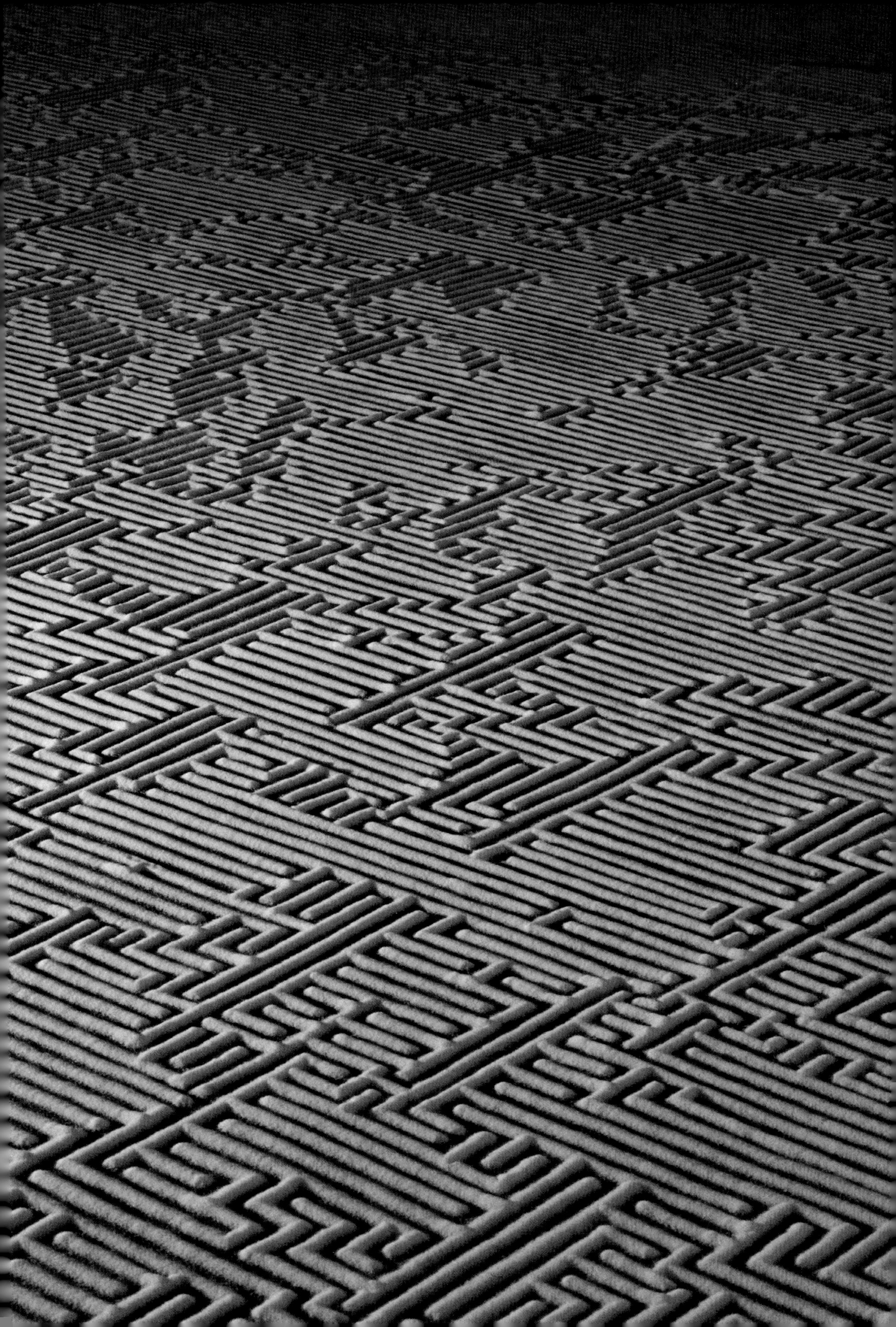

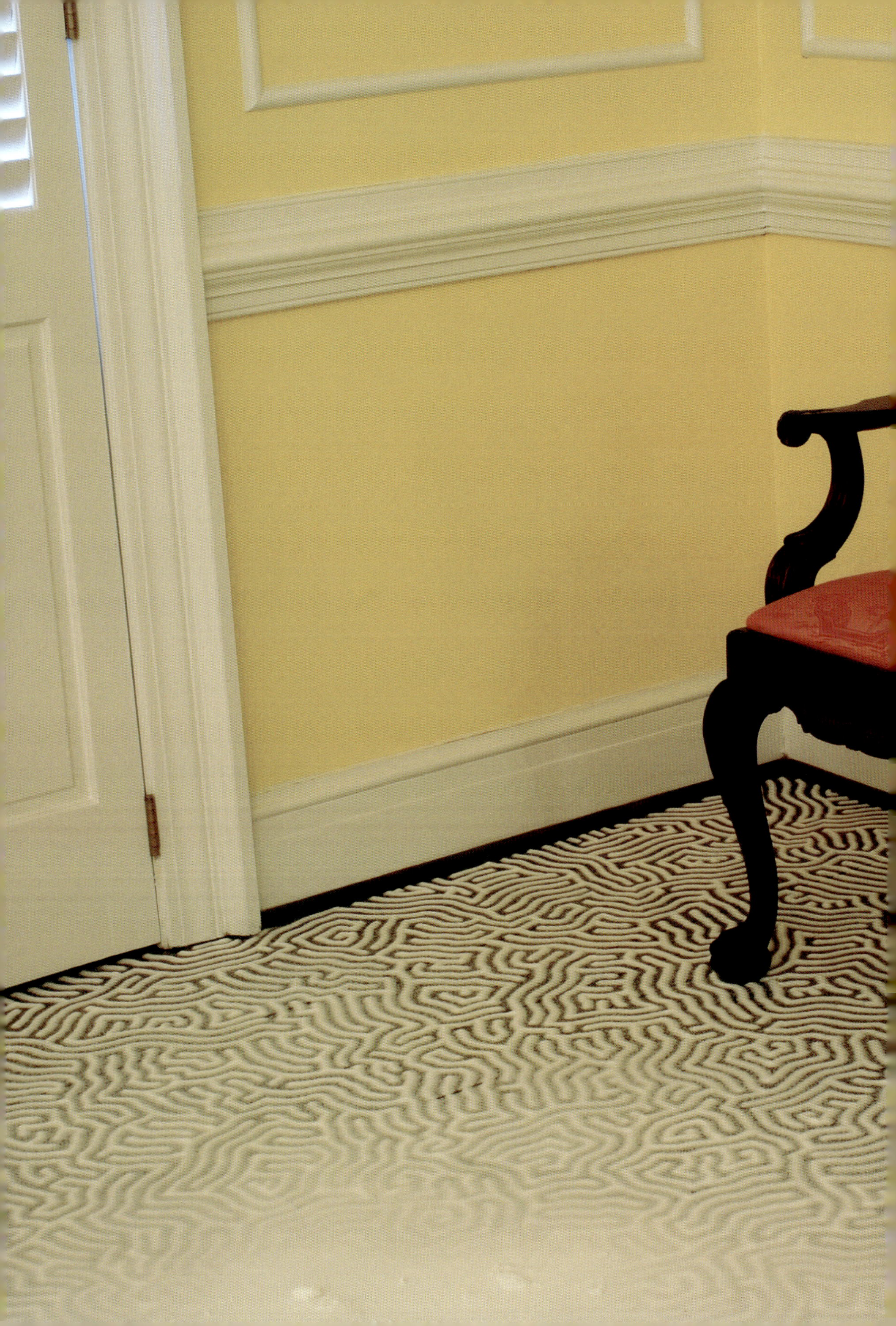

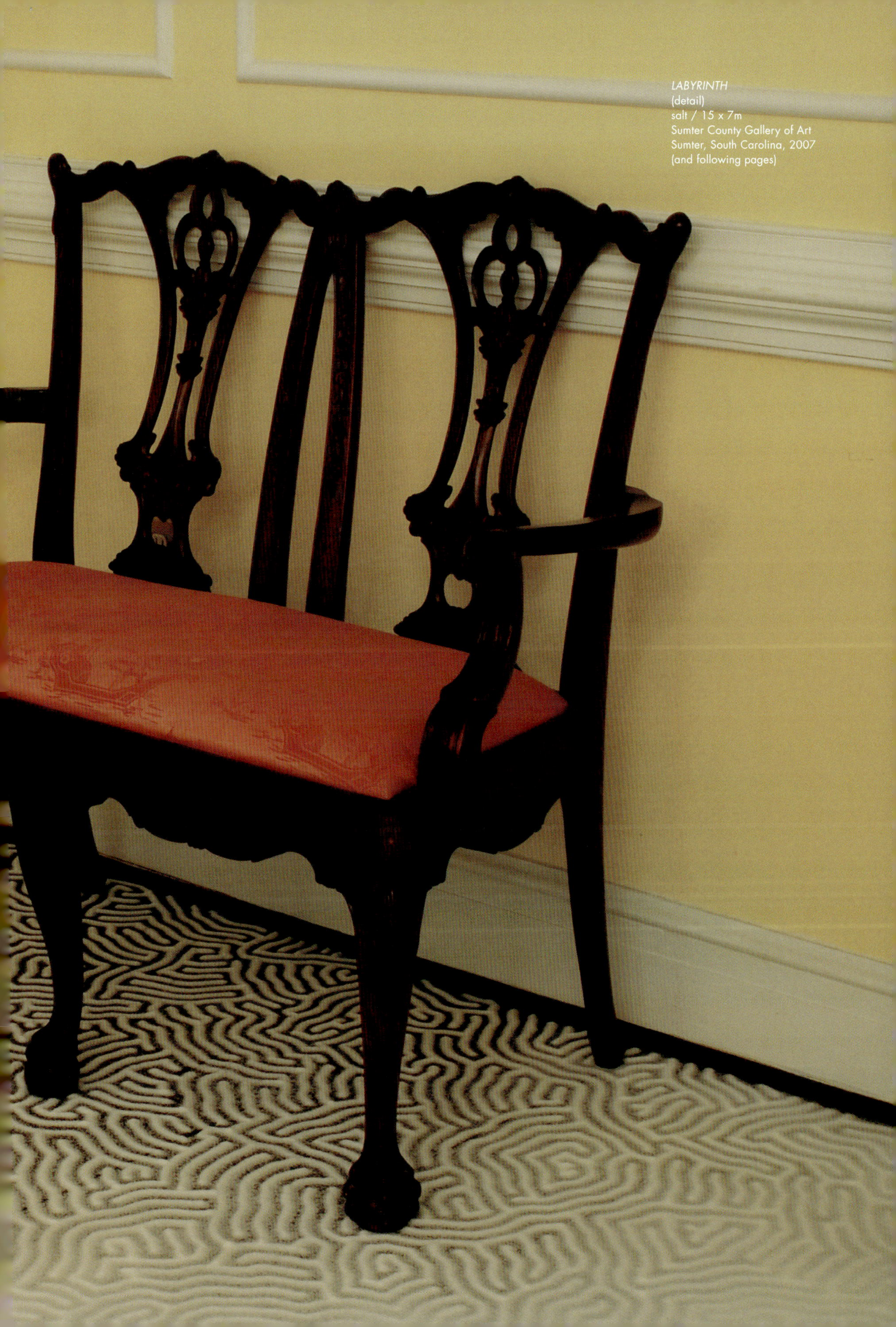

LABYRINTH
(detail)
salt / 15 x 7m
Sumter County Gallery of Art
Sumter, South Carolina, 2007
(and following pages)

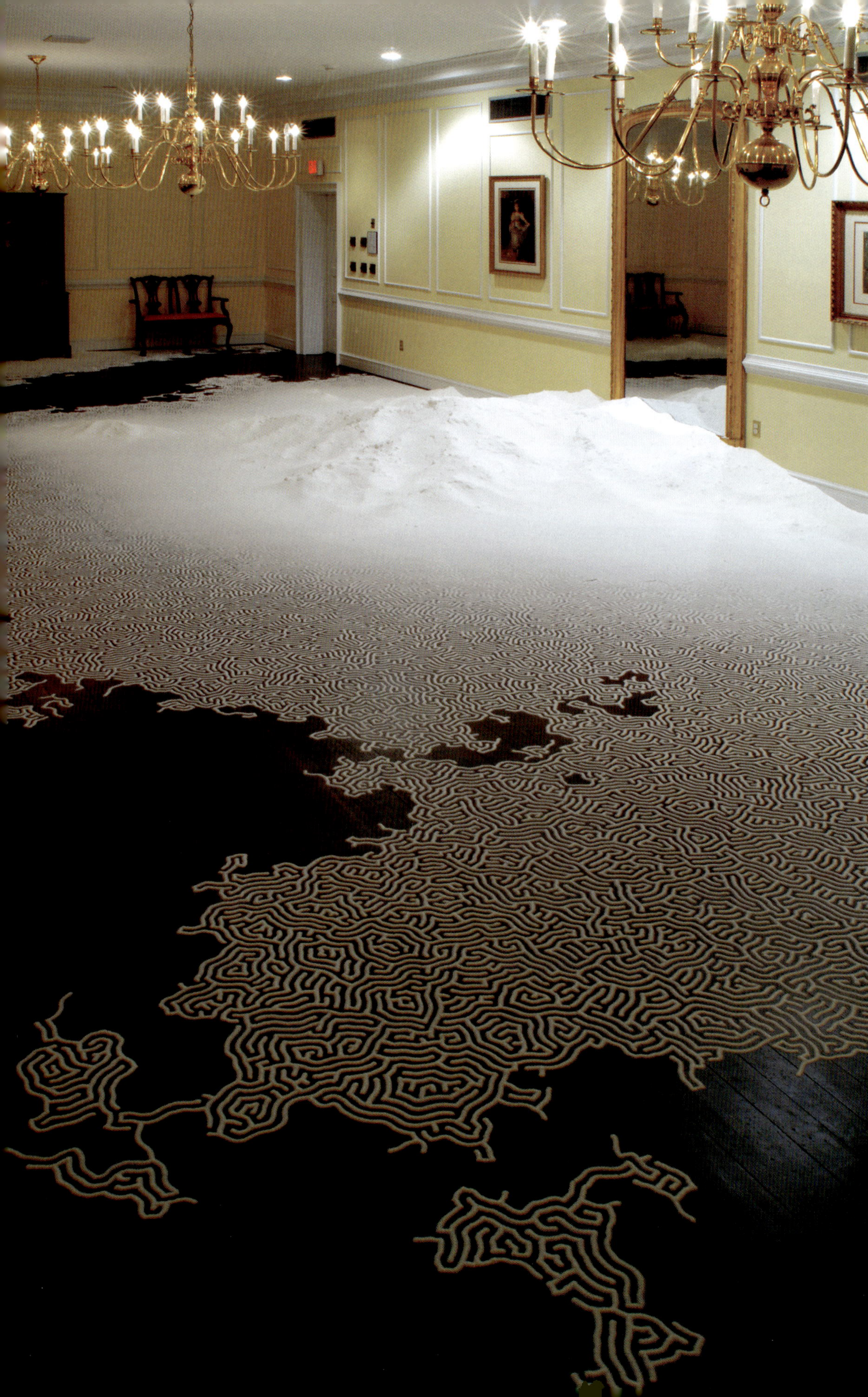

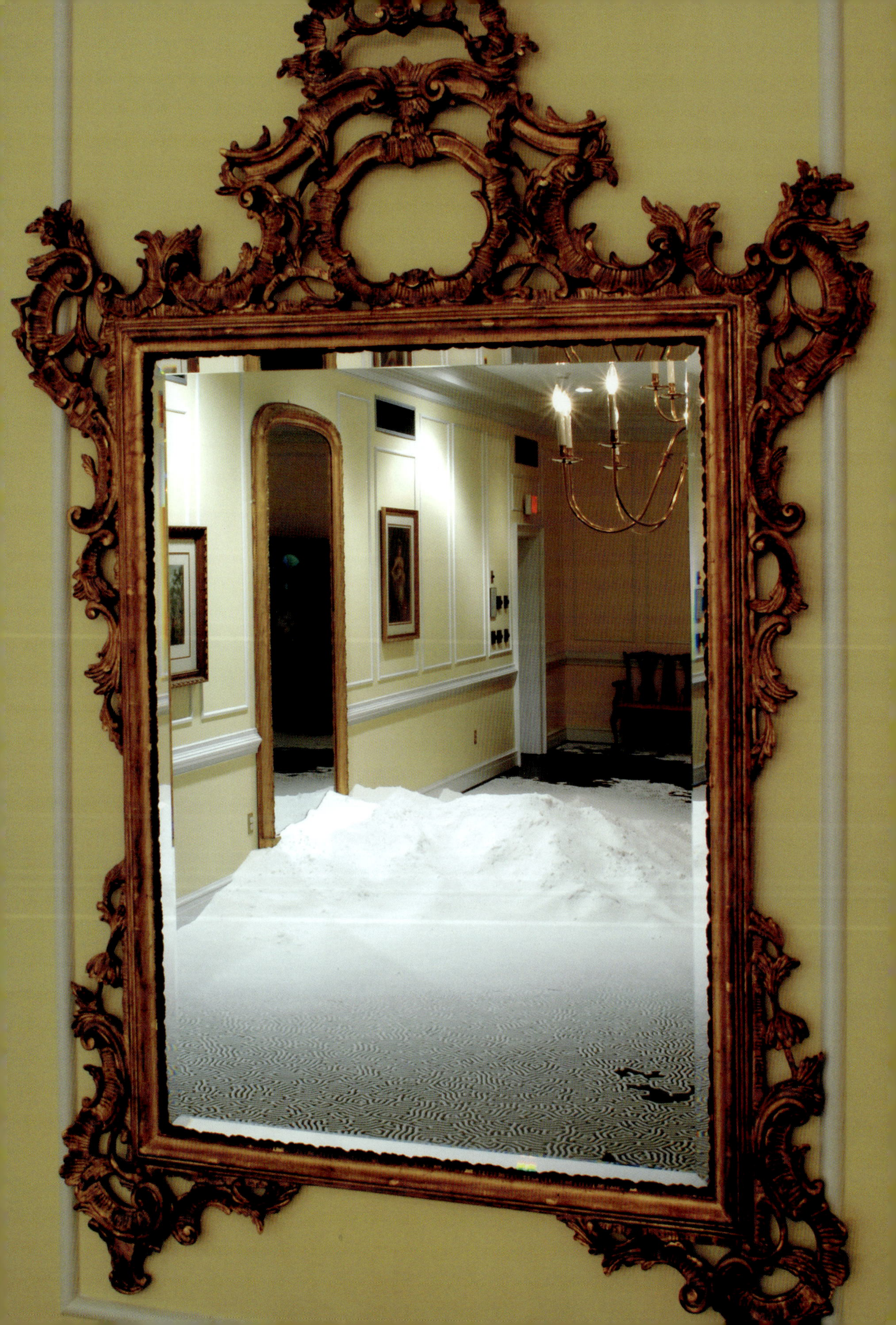

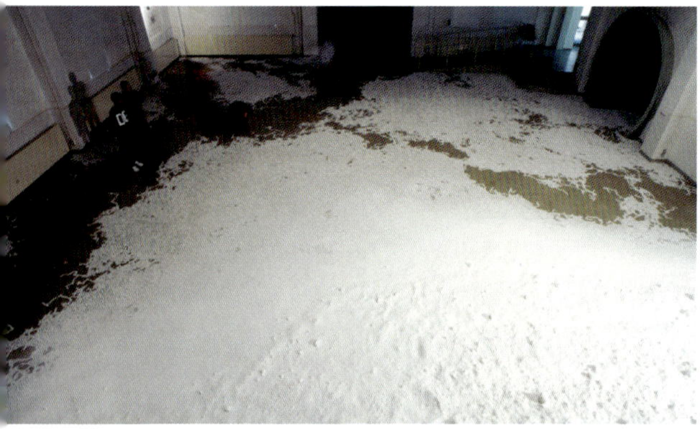

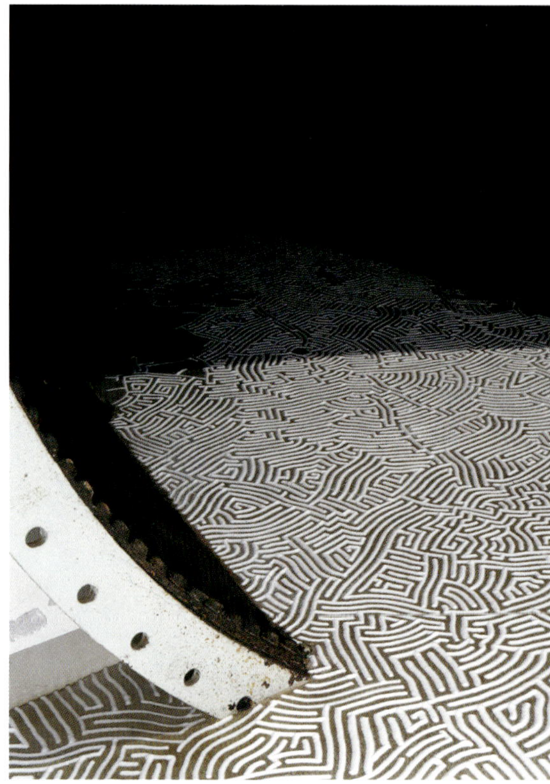

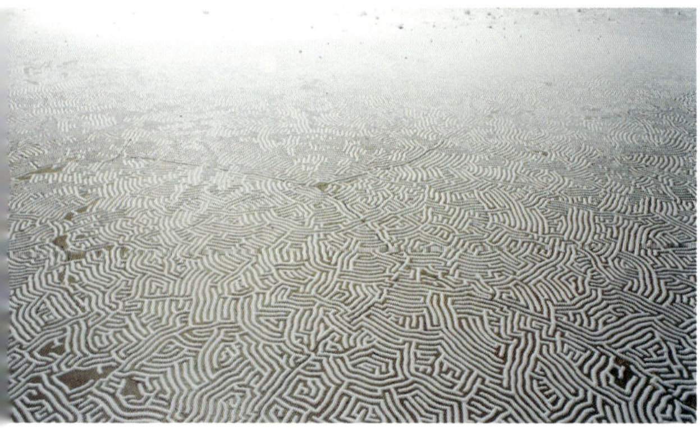

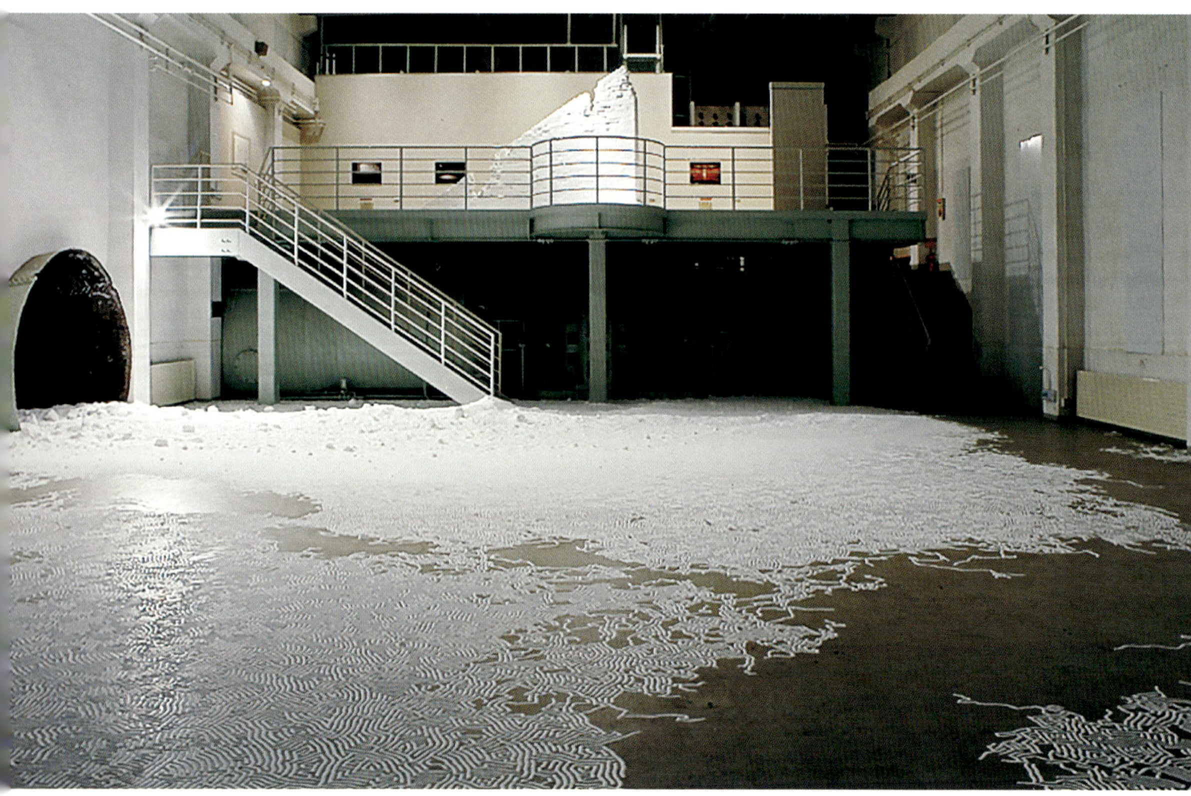

LABYRINTH
salt / 14.5 × 22.5m
Nizayama Forest Art Museum
Toyama, Japan, 2005

Photo: Sen Naganawa

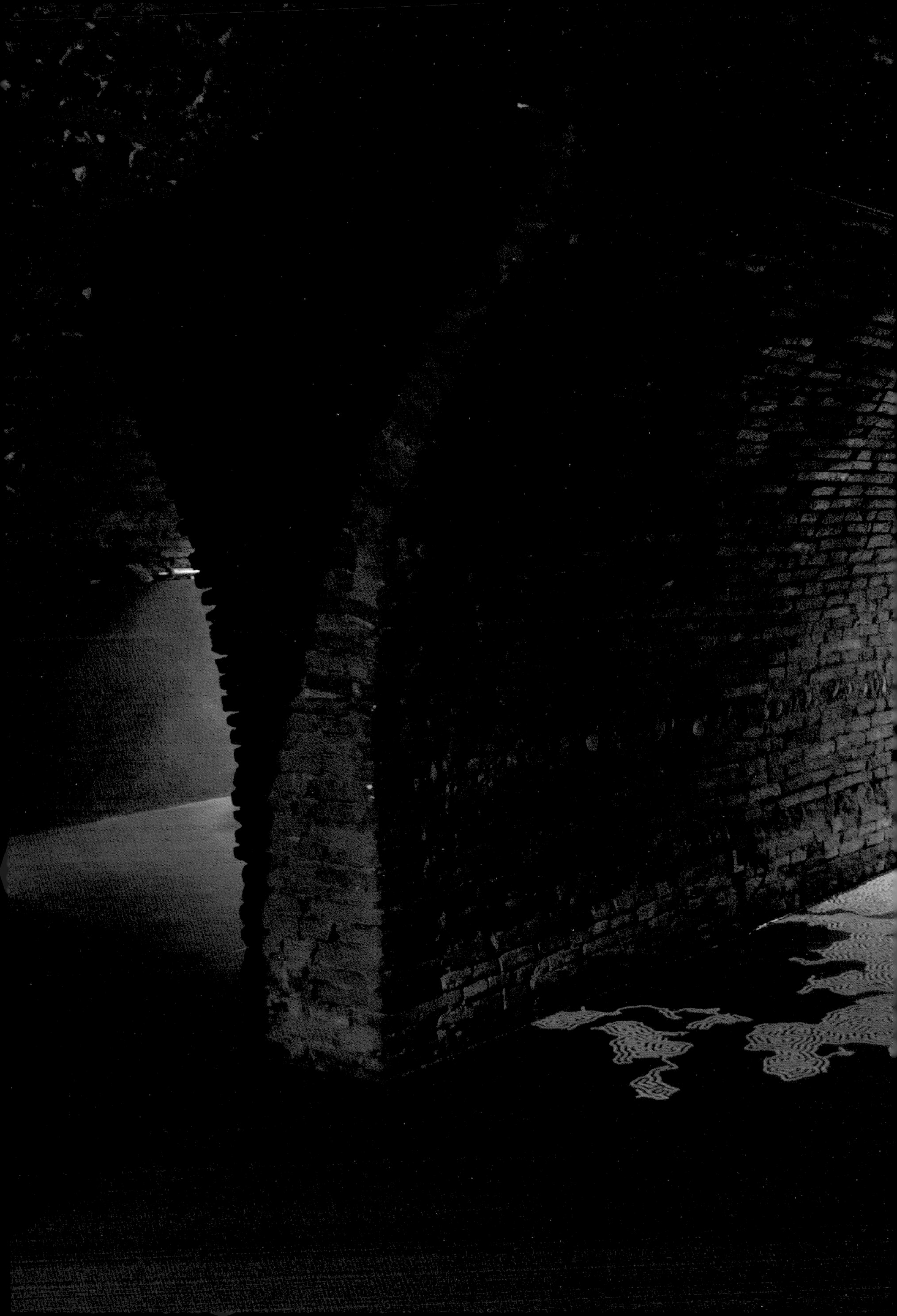

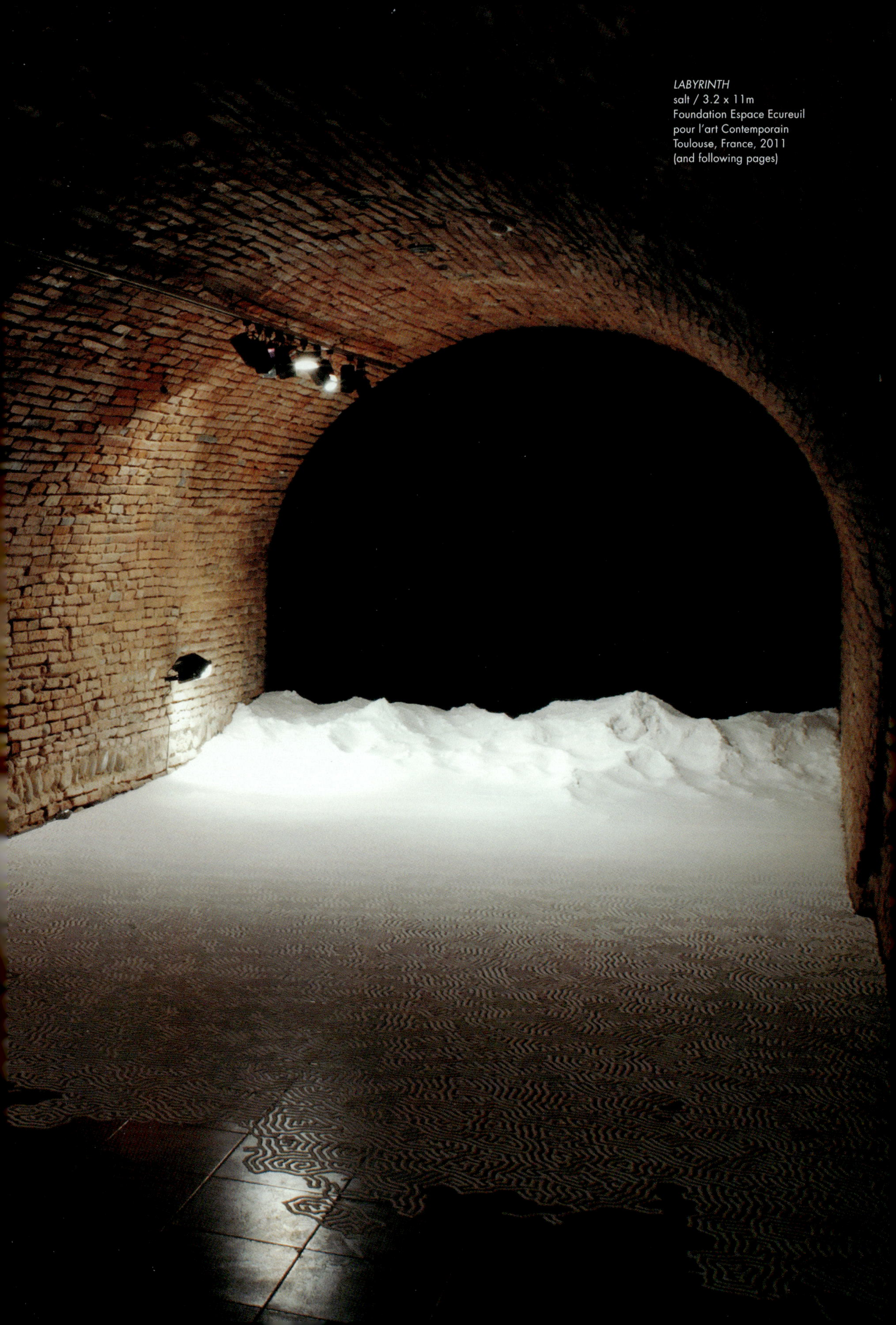

LABYRINTH
salt / 3.2 x 11m
Foundation Espace Ecureuil
pour l'art Contemporain
Toulouse, France, 2011
(and following pages)

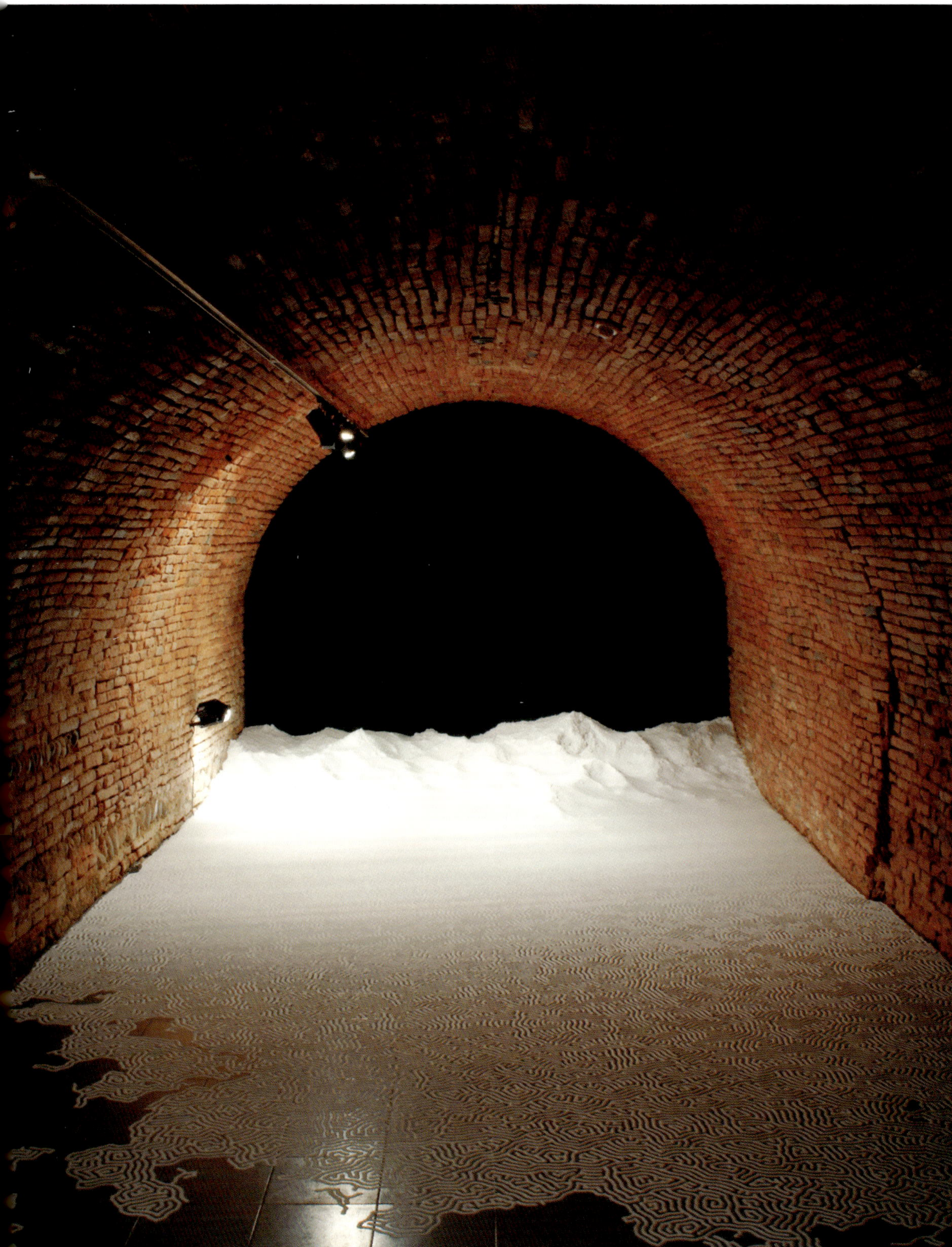

LABYRINTH
salt / 7.2 × 7.2m
Addlestone Library
College of Charleston
Halsey Institute of Contemporary Art
Charleston, South Carolina, 2006
(and following pages)

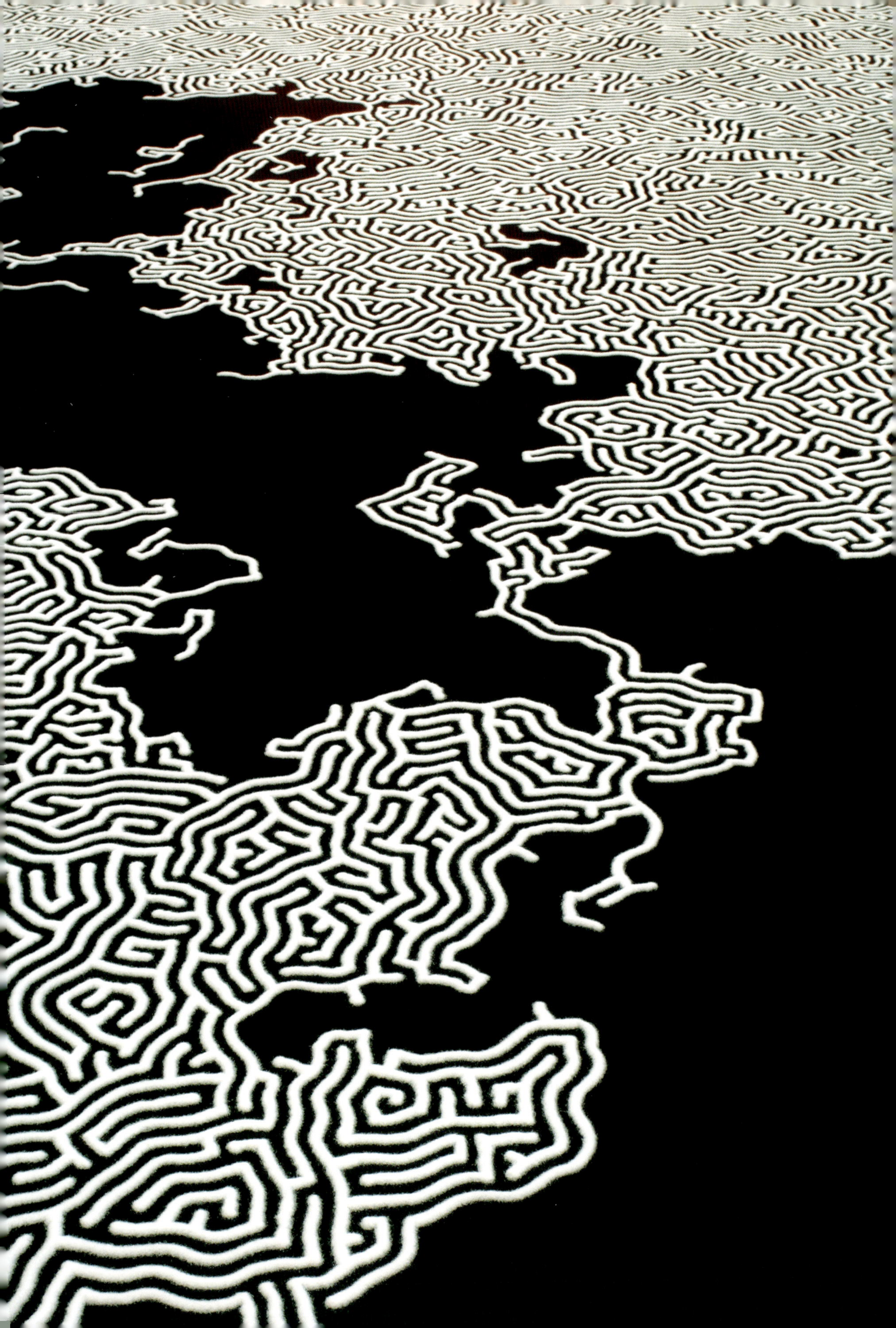

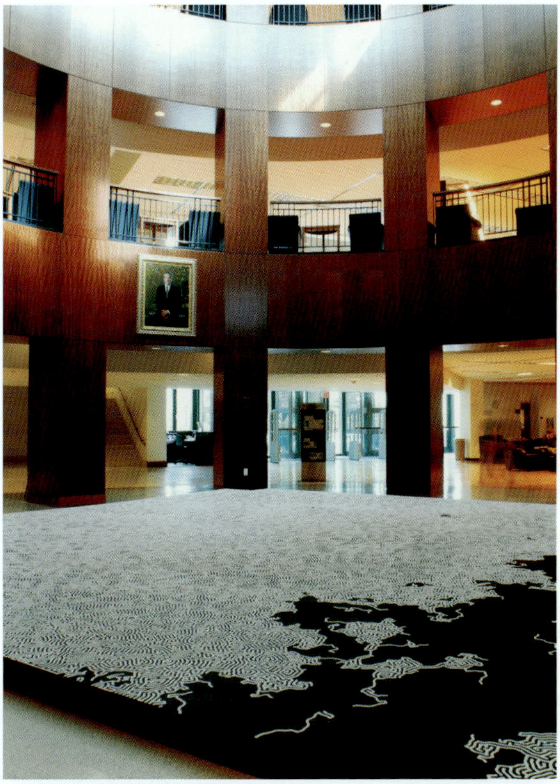

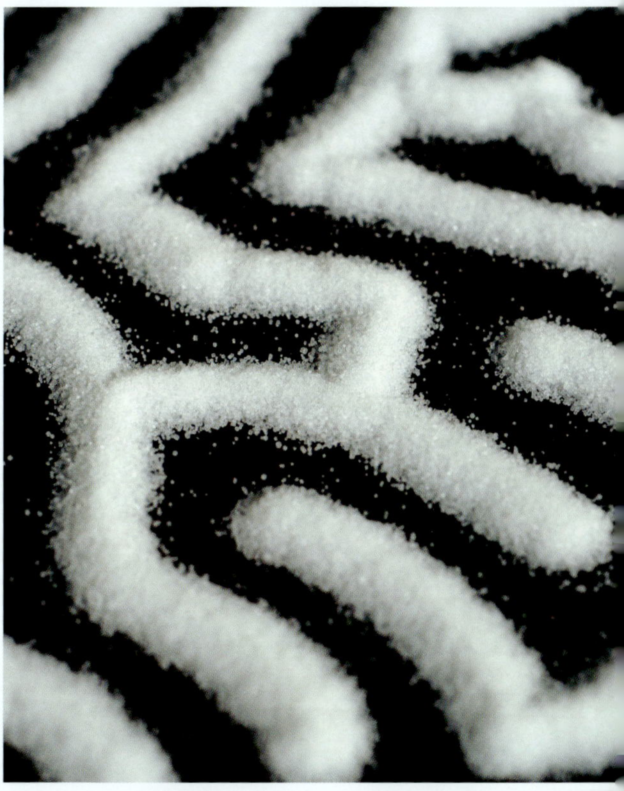

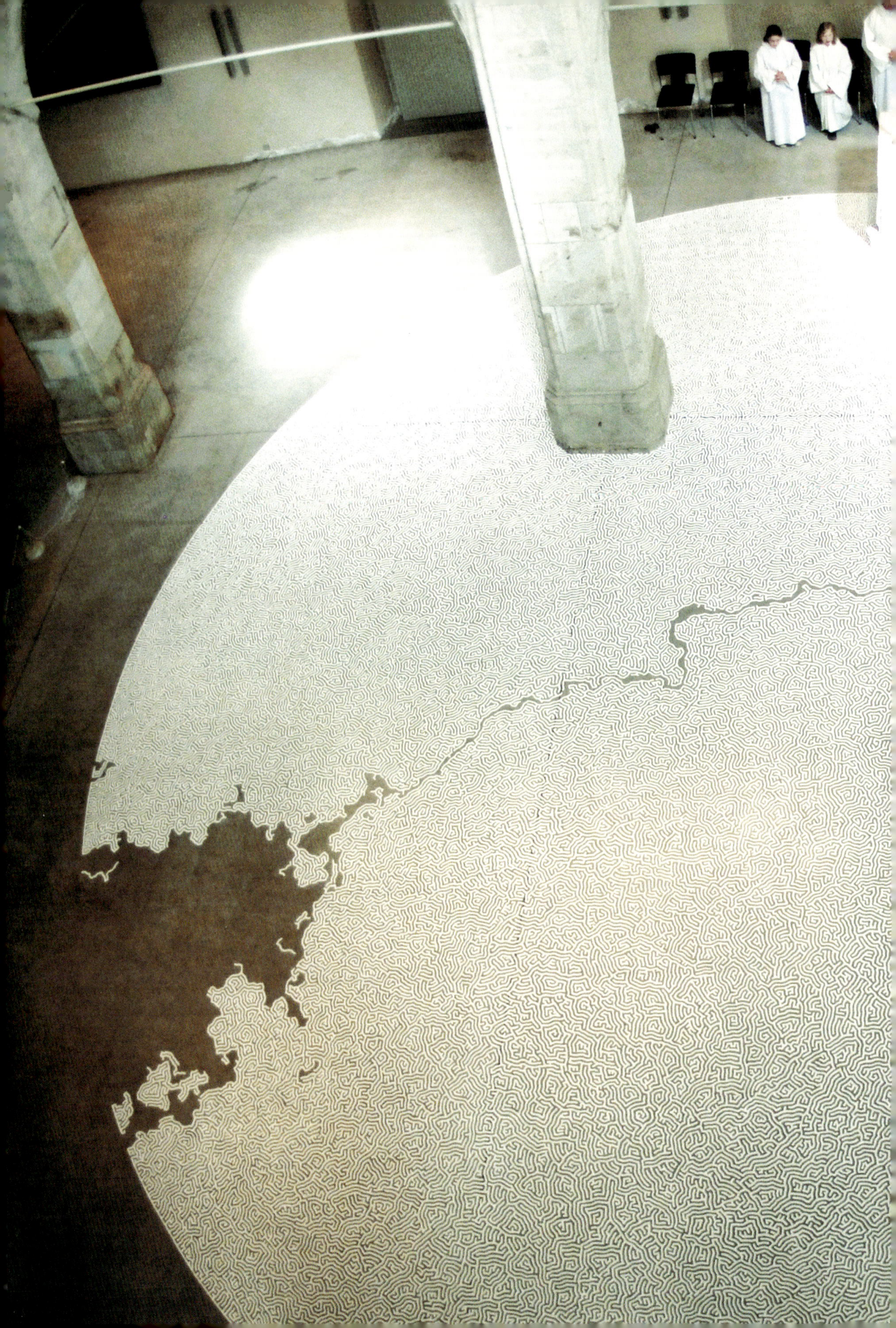

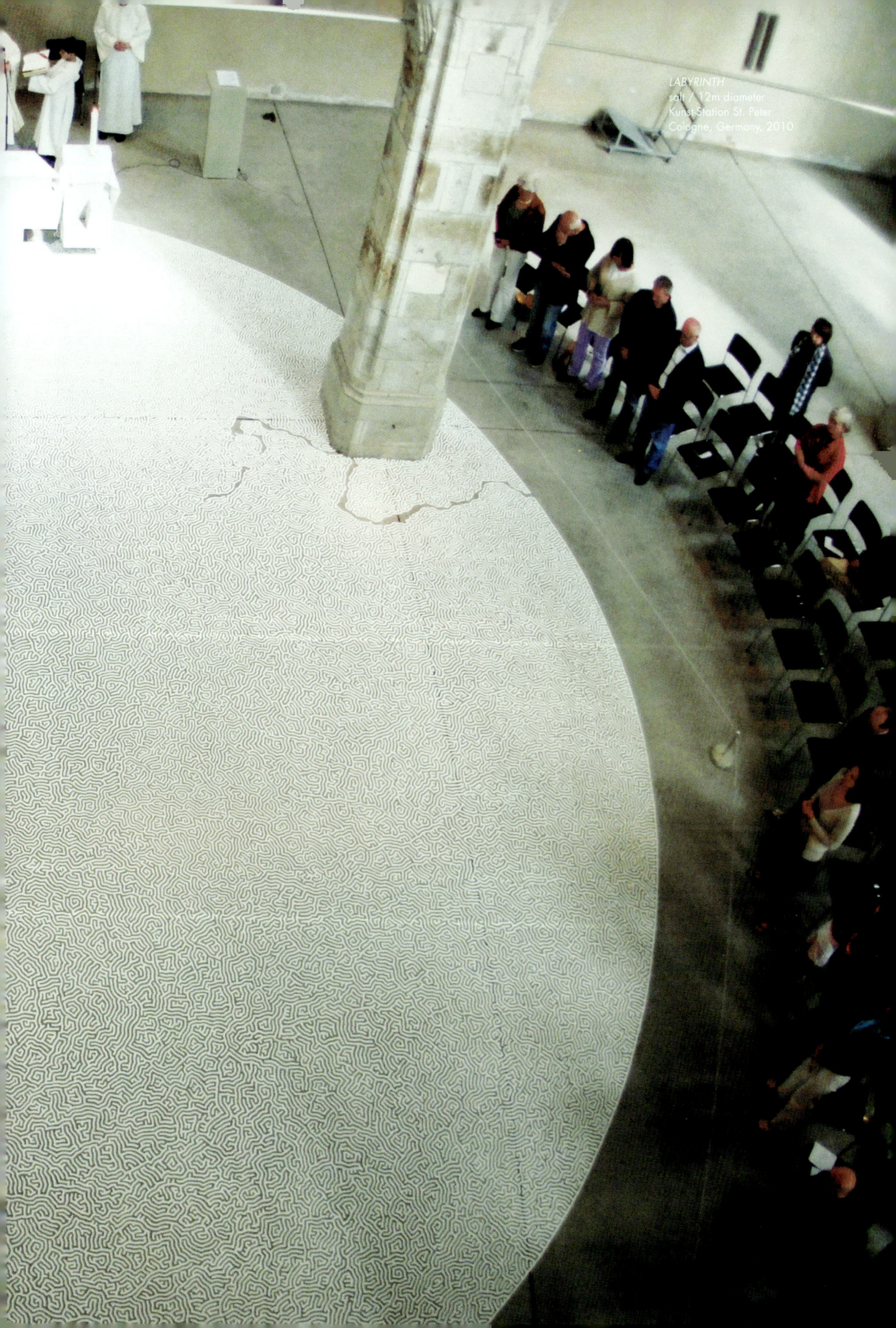

LABYRINTH
salt / 12m diameter
Kunst-Station St. Peter
Cologne, Germany, 2010

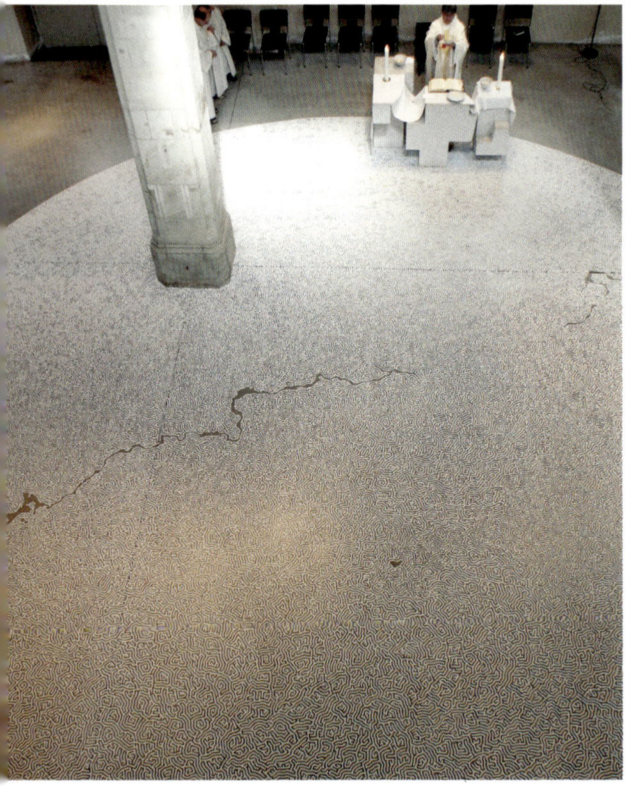
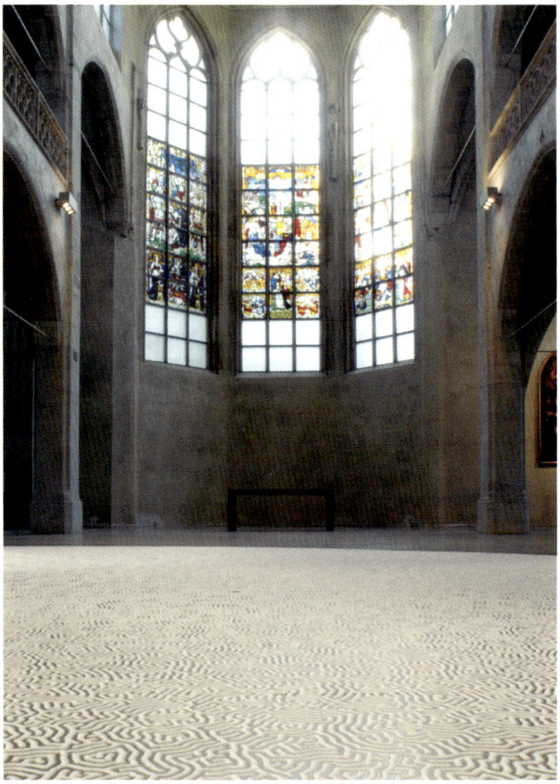
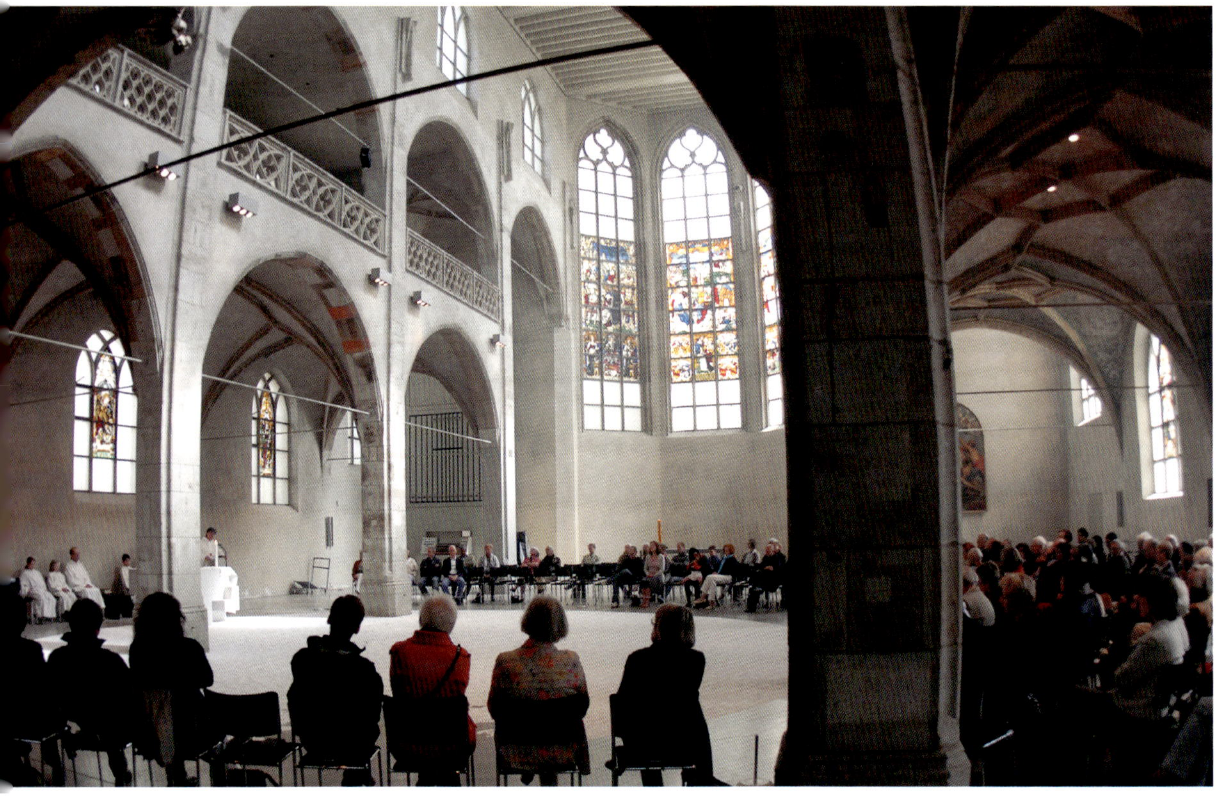

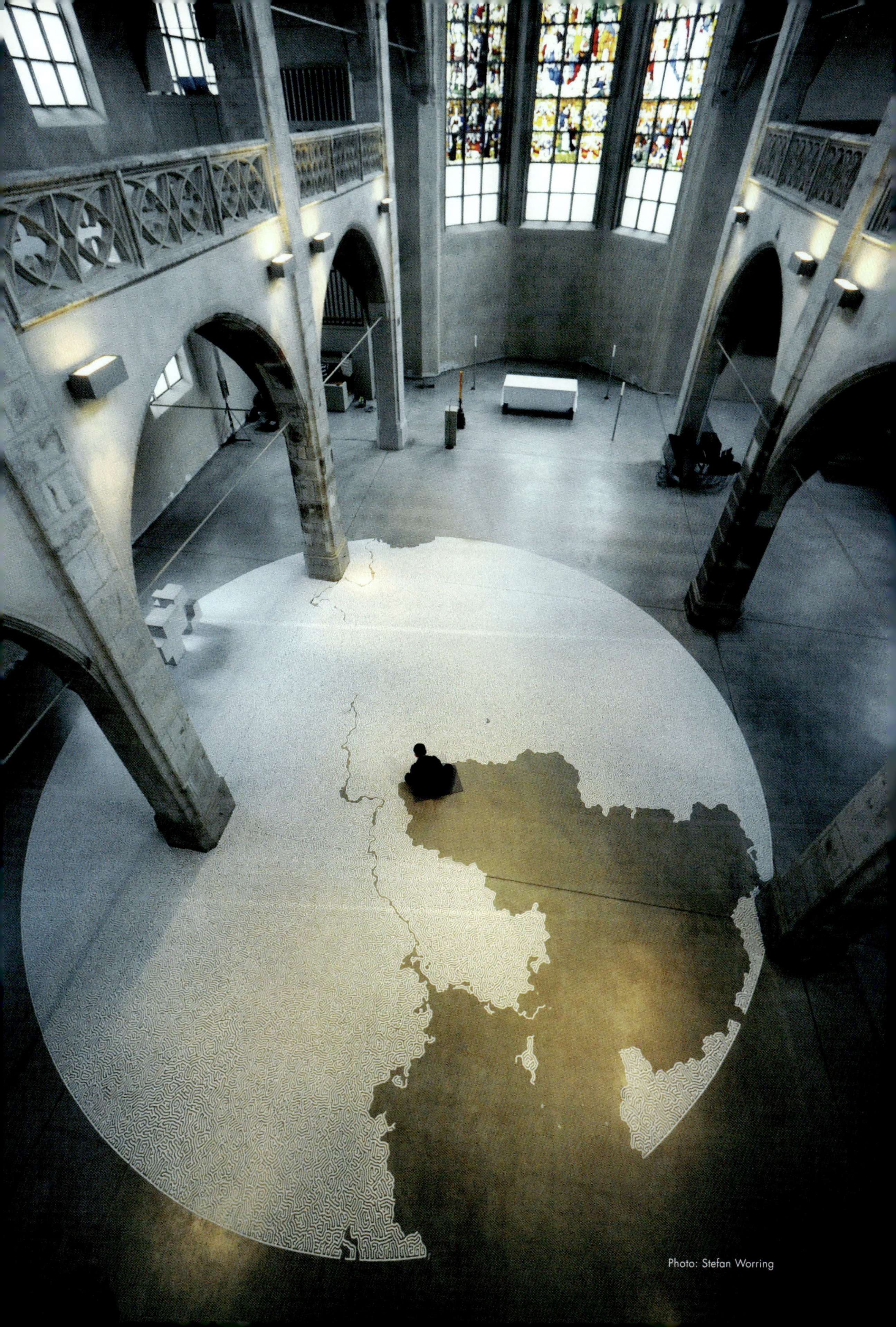

Photo: Stefan Worring

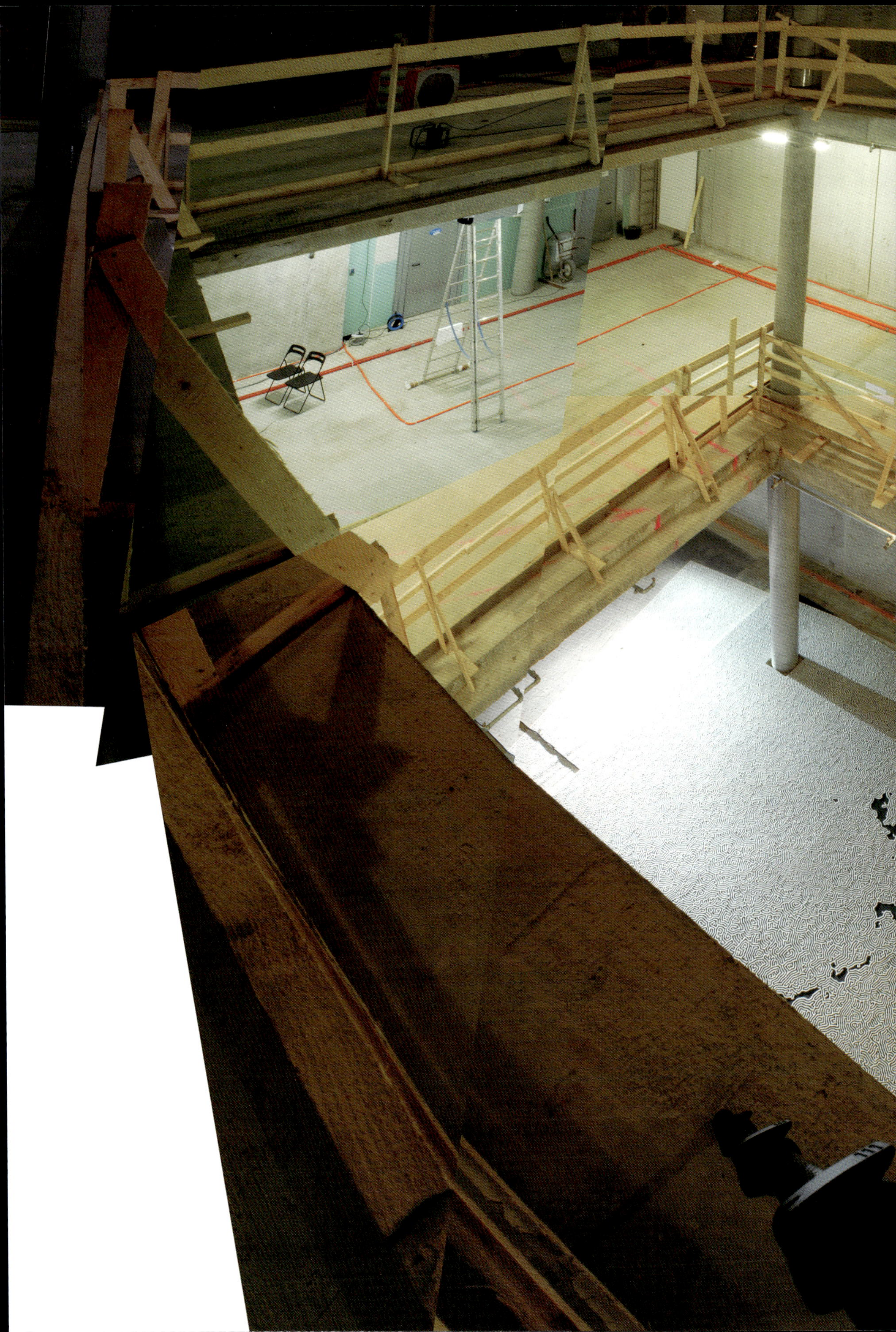

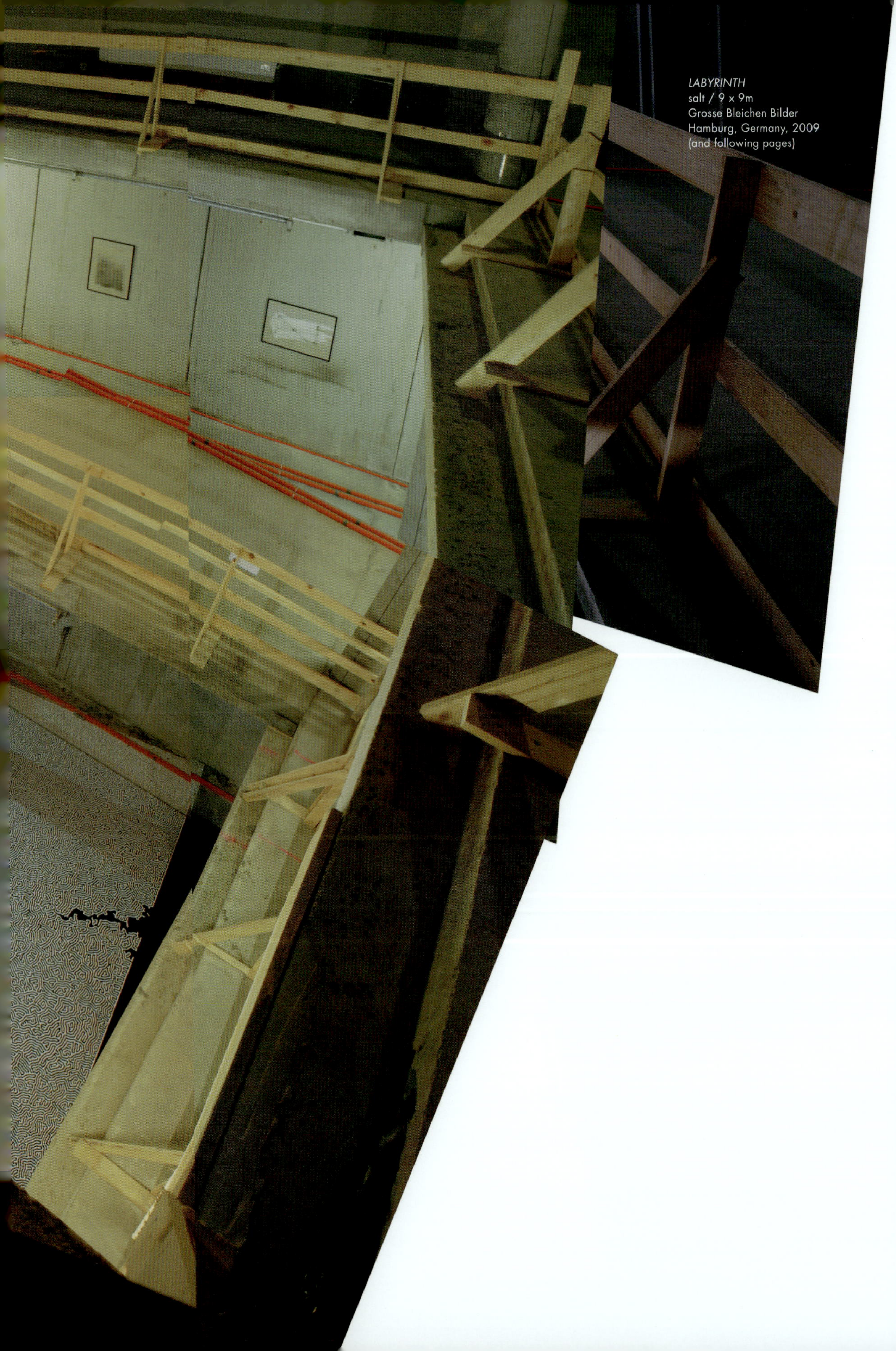

LABYRINTH
salt / 9 x 9m
Grosse Bleichen Bilder
Hamburg, Germany, 2009
(and following pages)

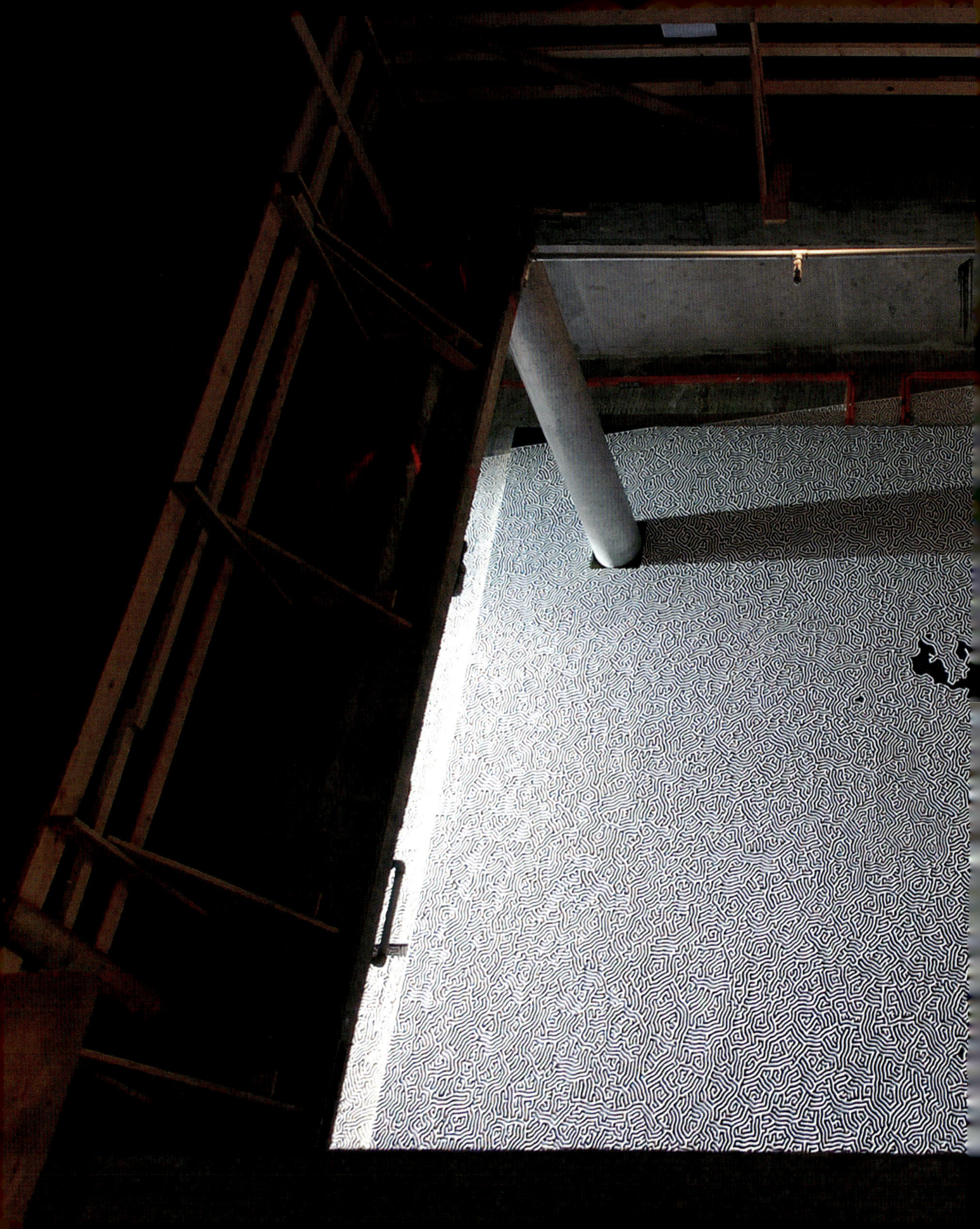

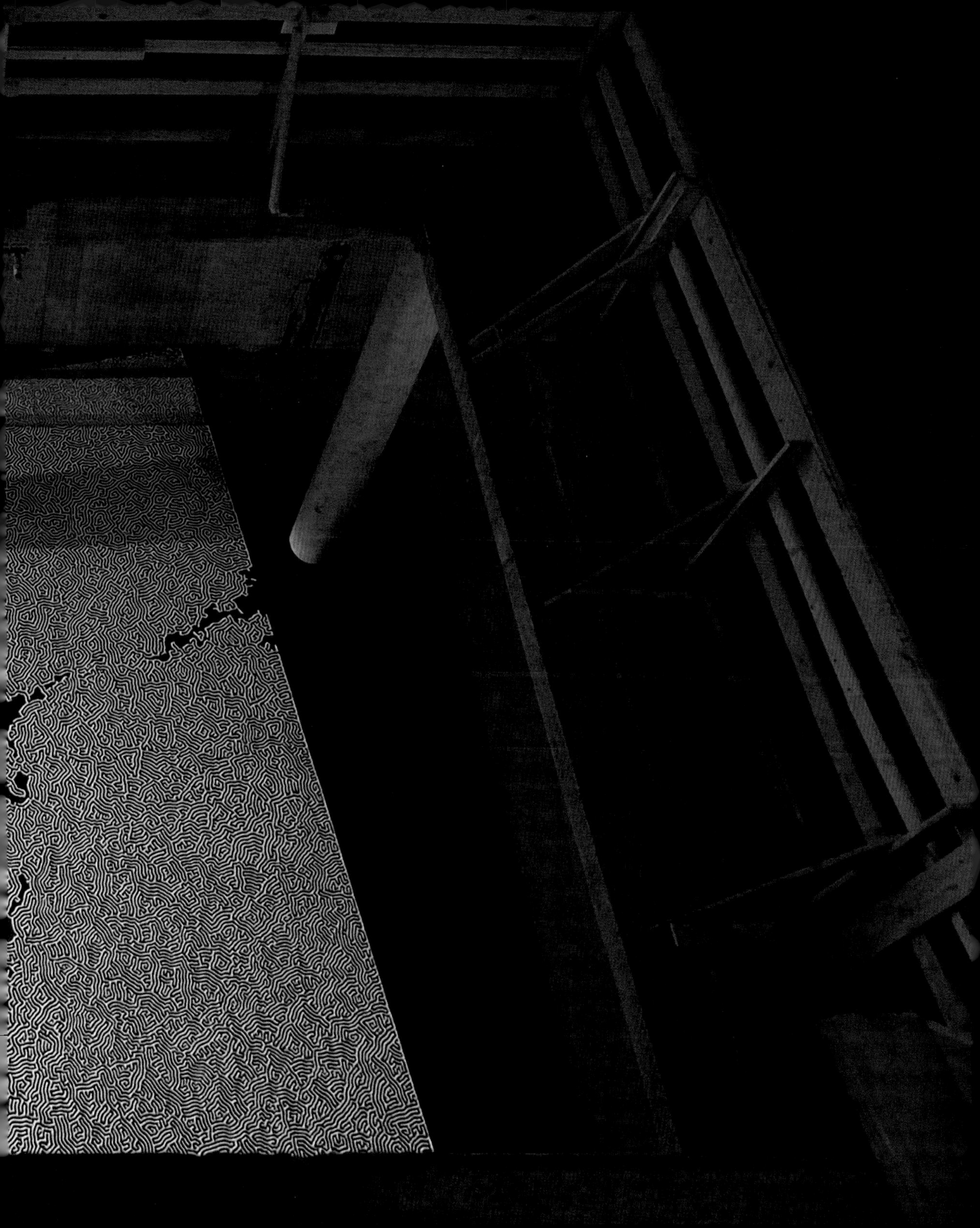

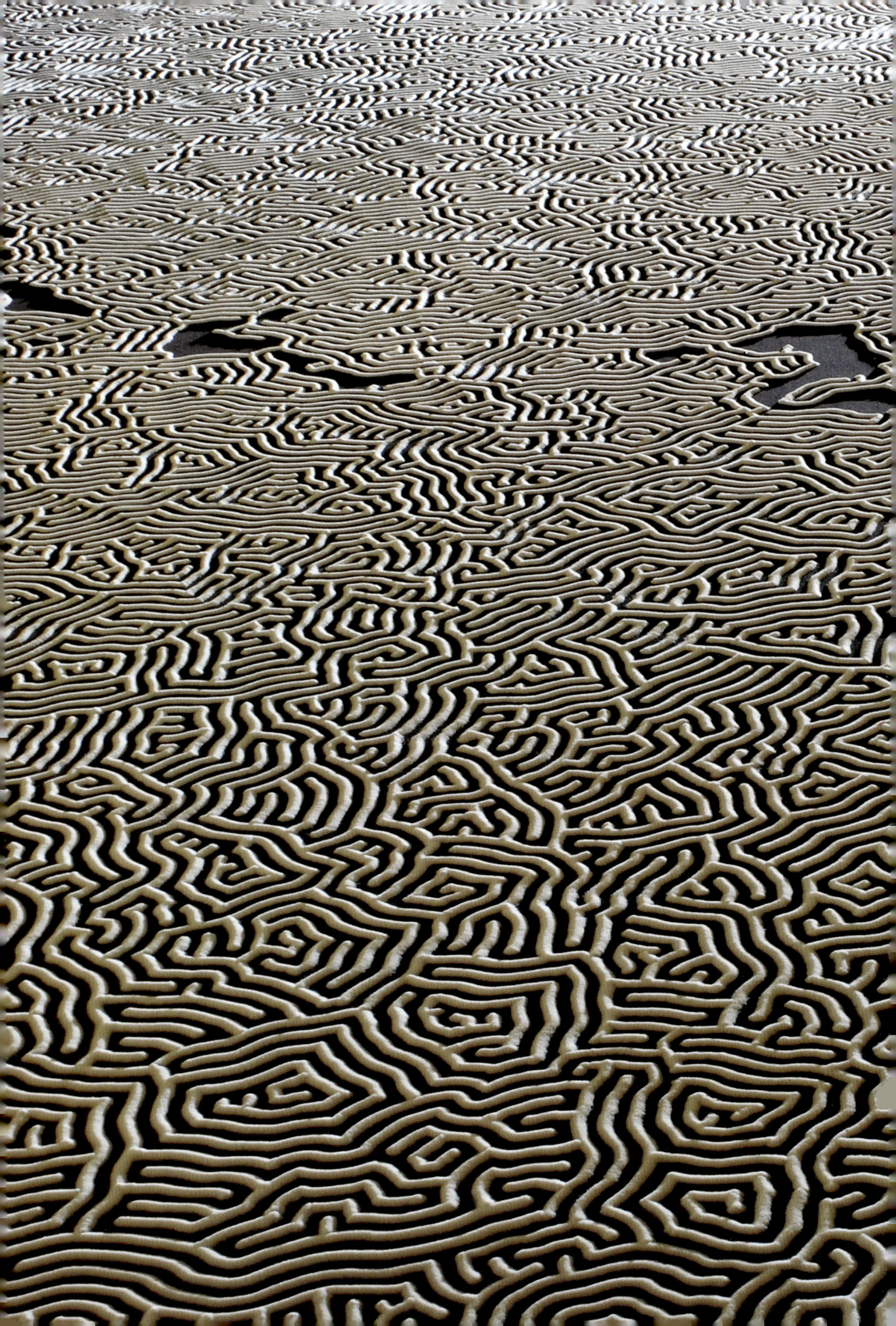

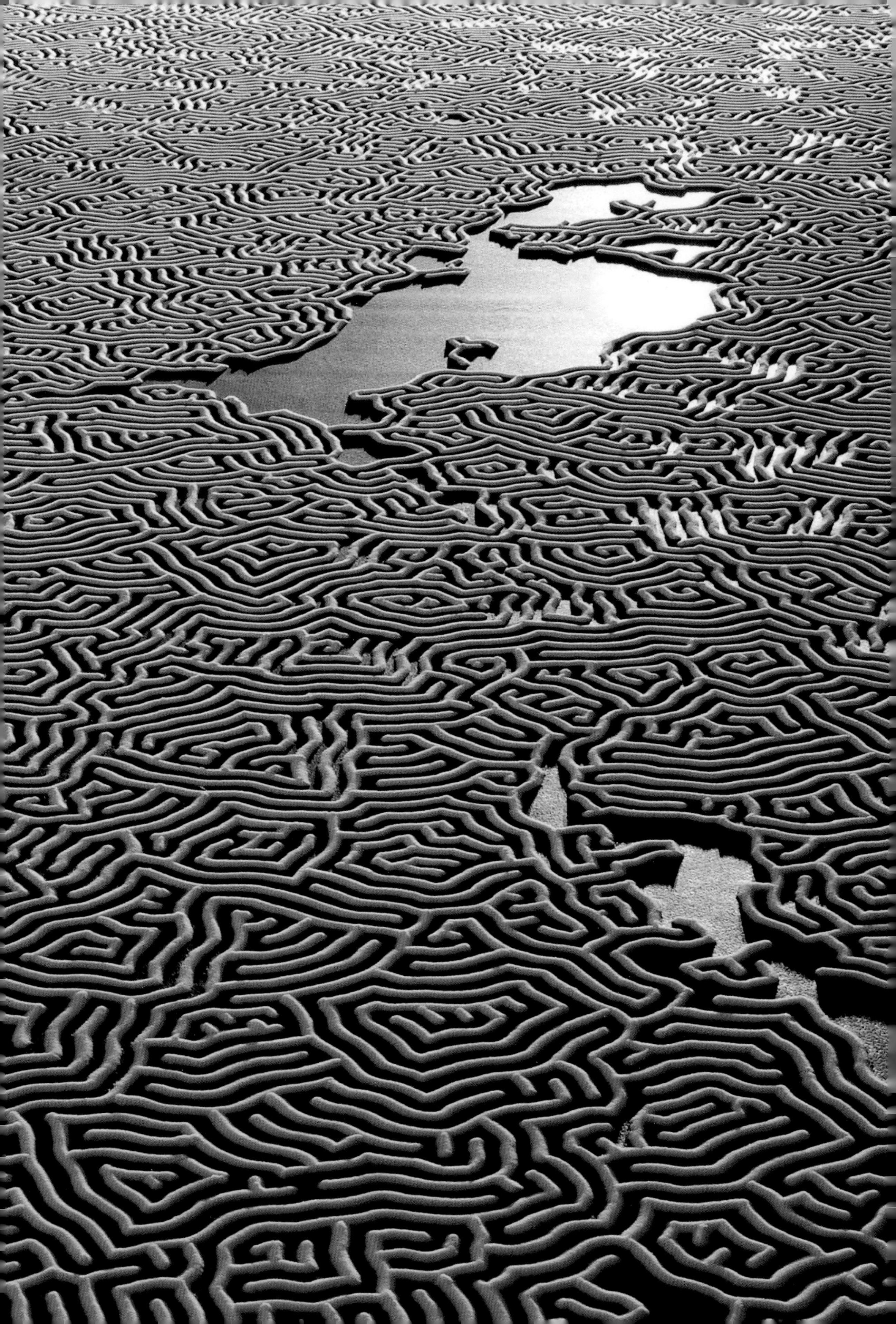

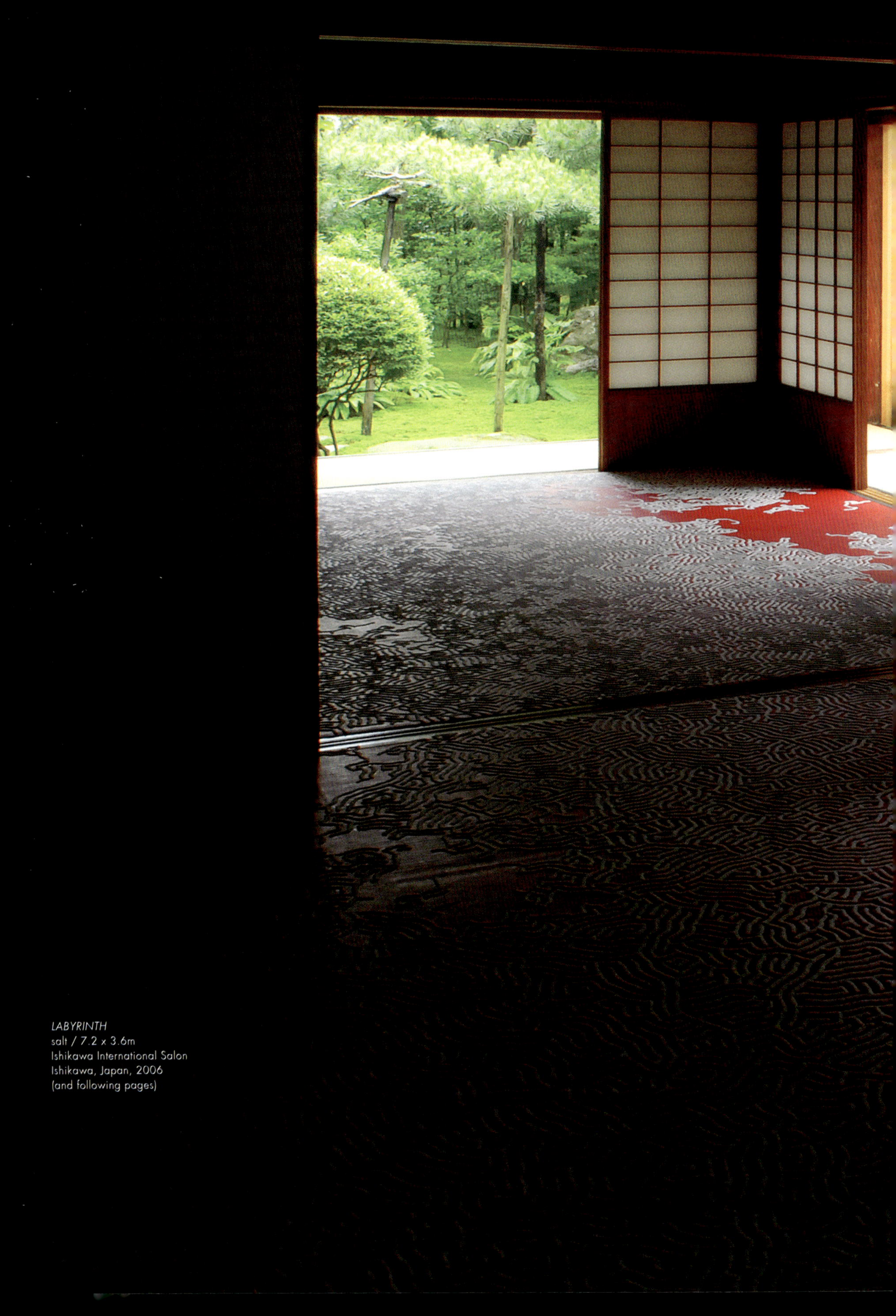

LABYRINTH
salt / 7.2 x 3.6m
Ishikawa International Salon
Ishikawa, Japan, 2006
(and following pages)

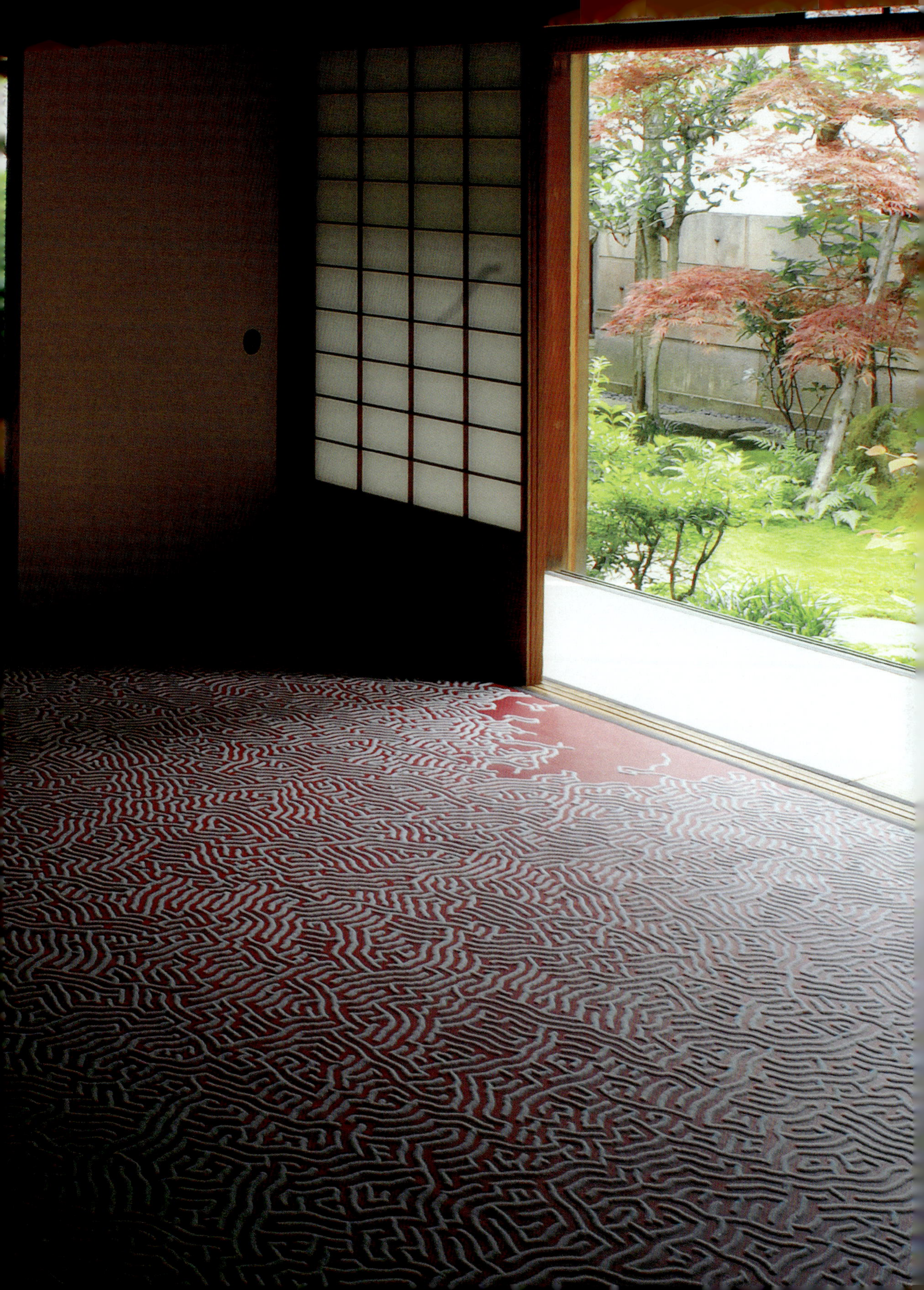

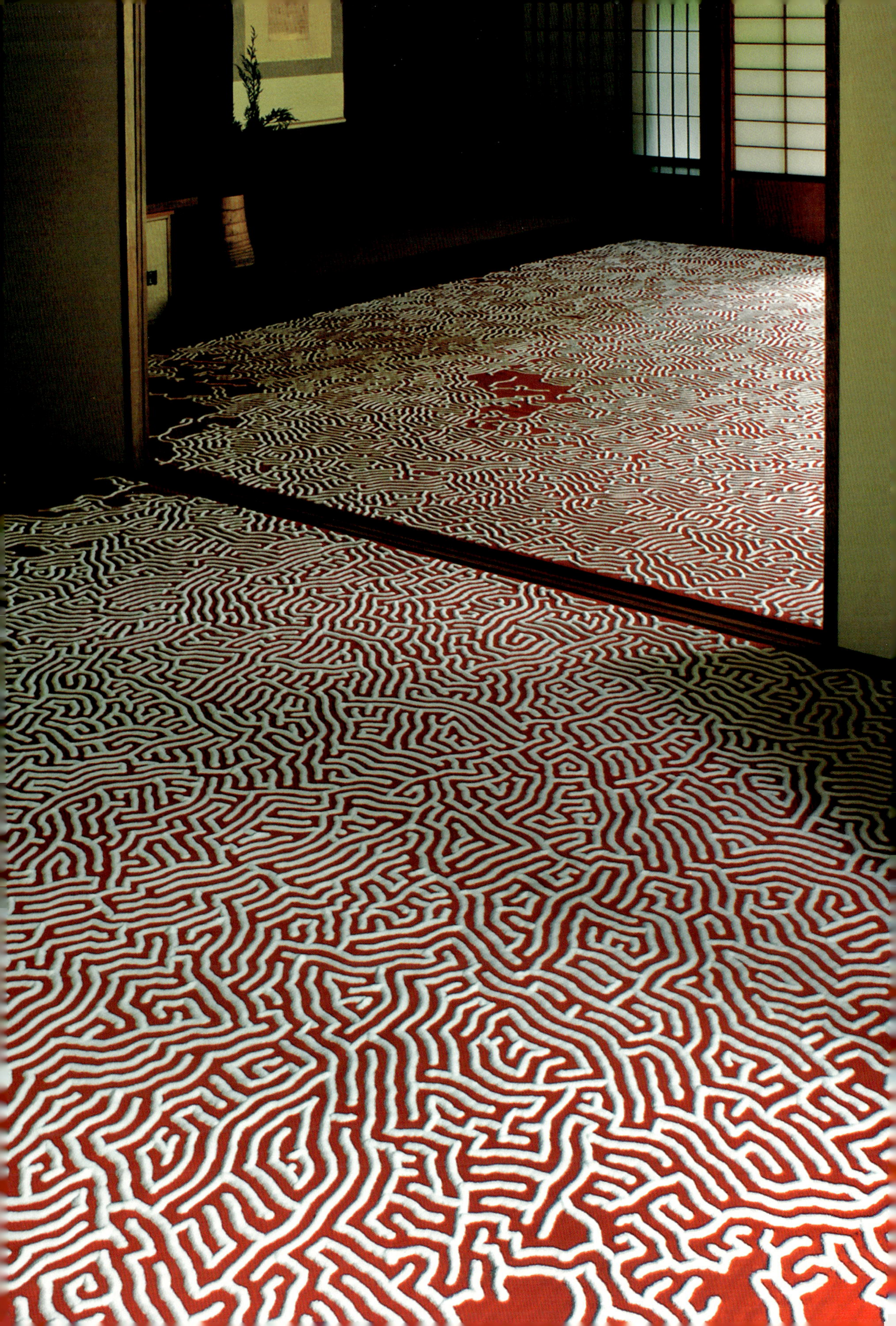

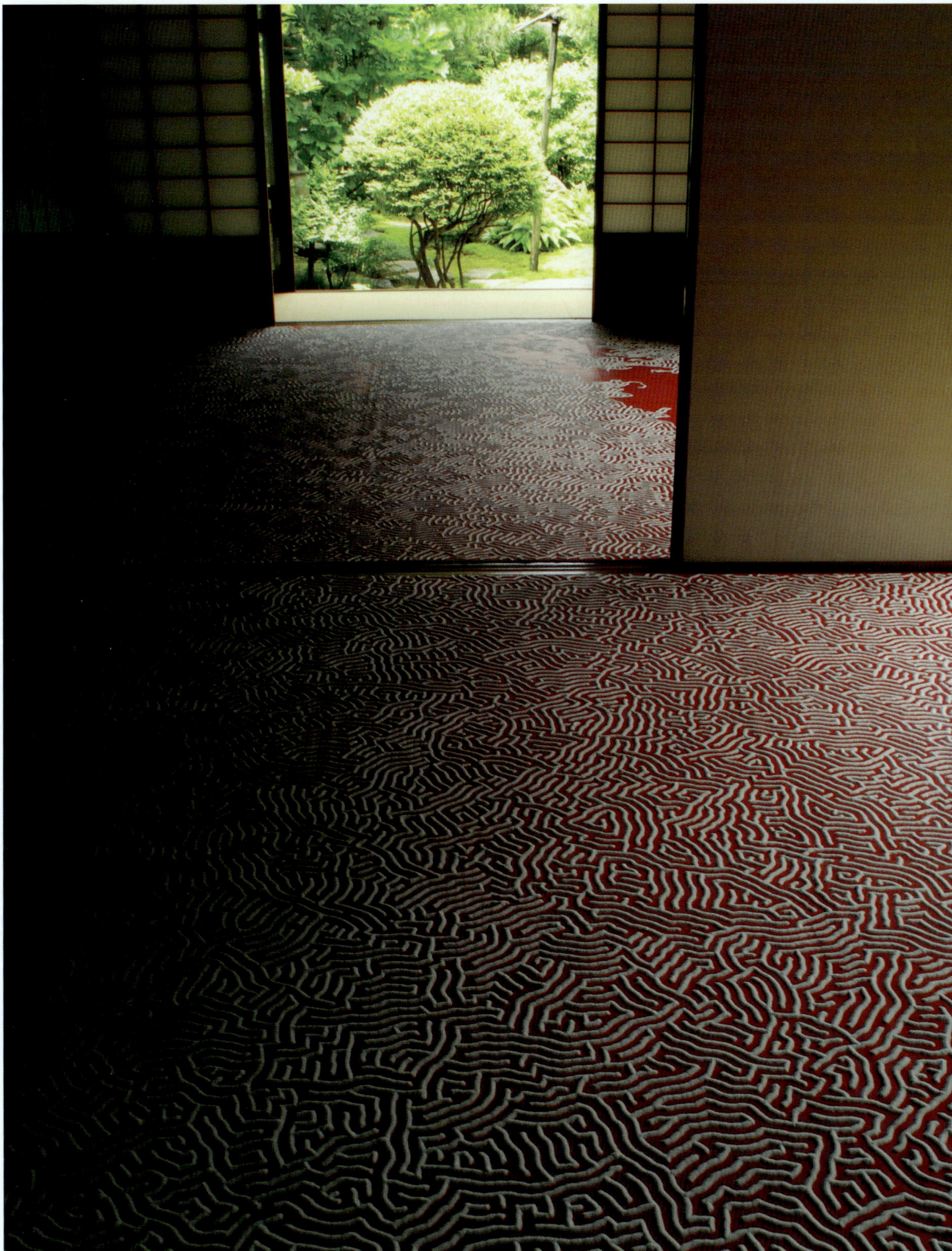

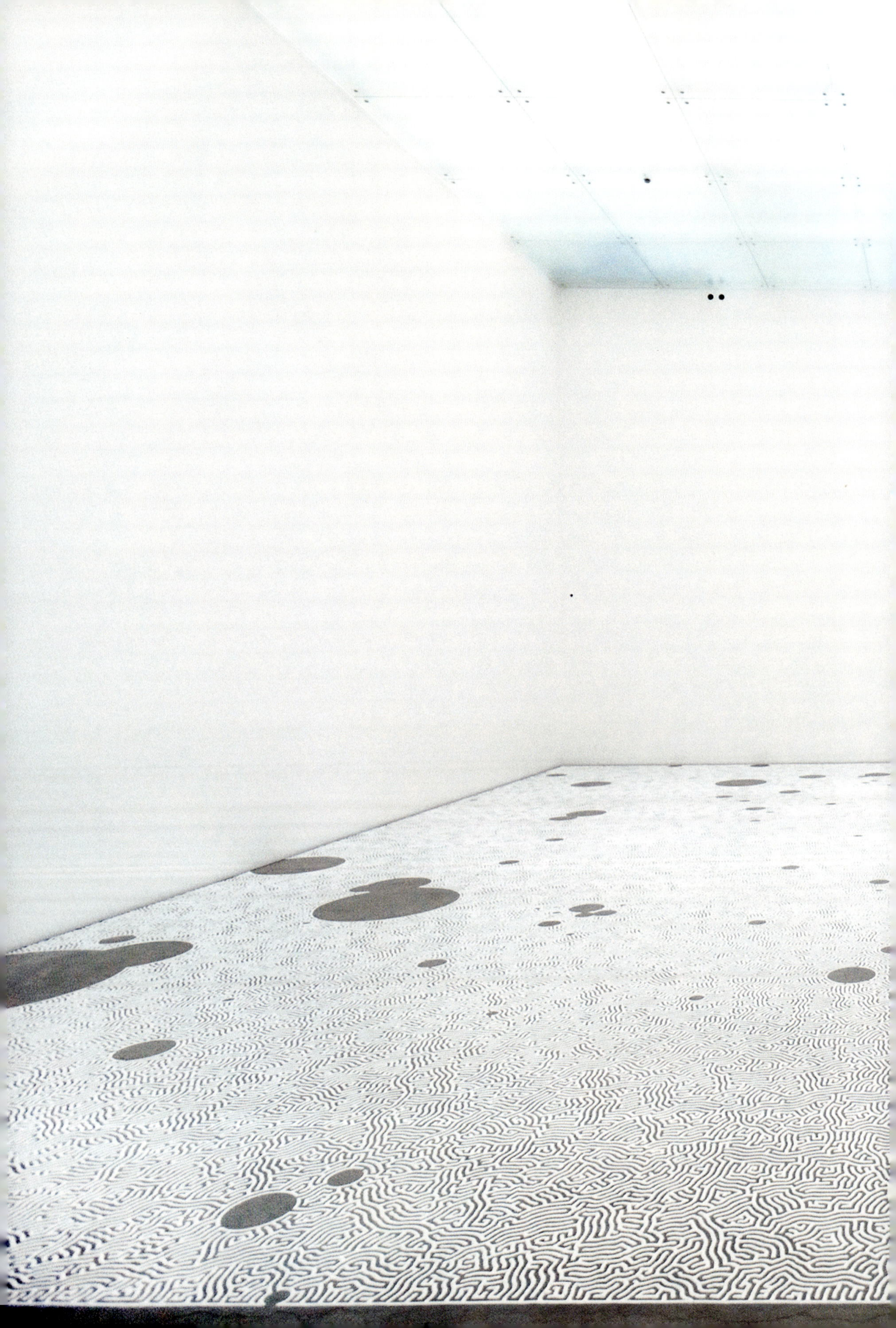

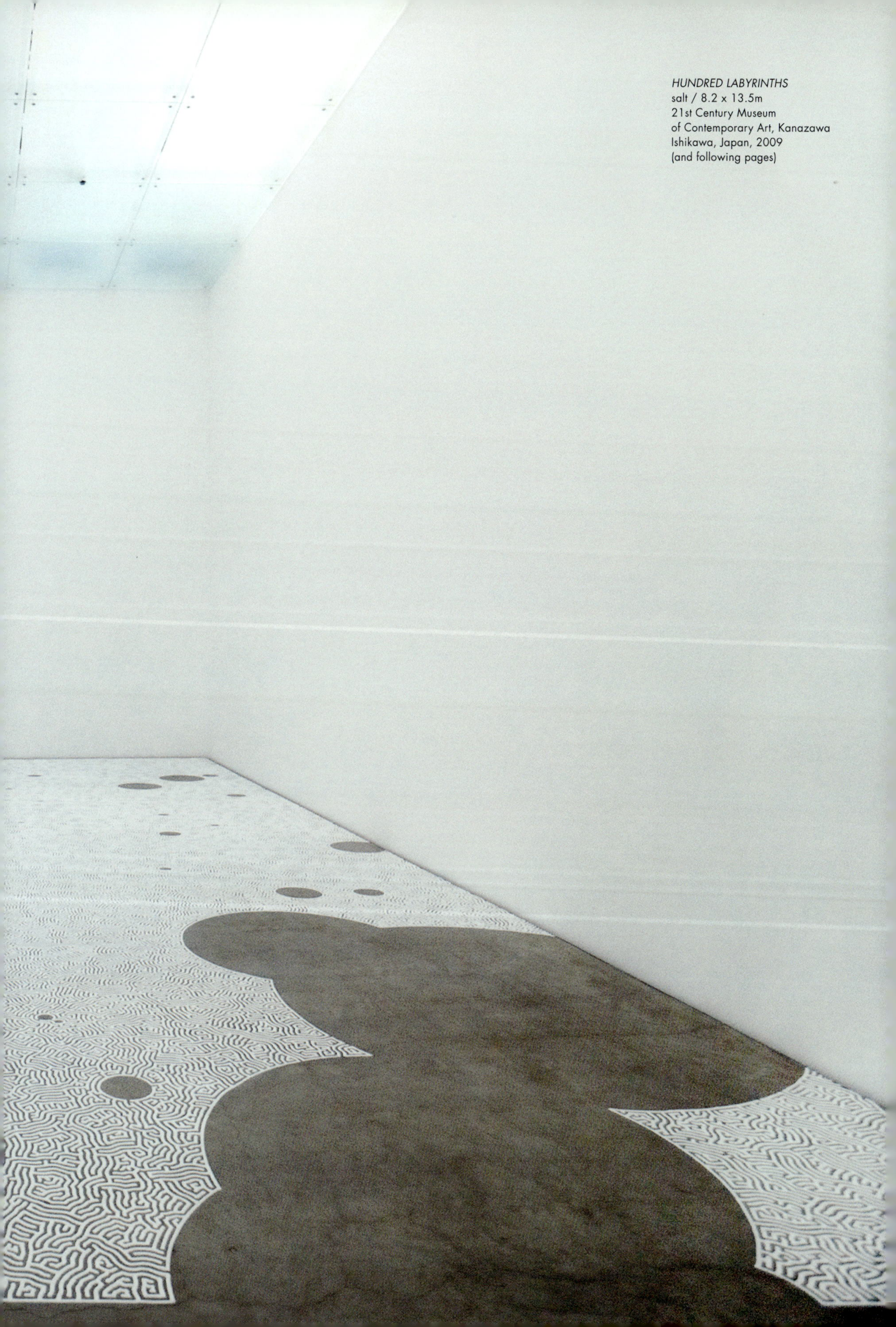

HUNDRED LABYRINTHS
salt / 8.2 x 13.5m
21st Century Museum
of Contemporary Art, Kanazawa
Ishikawa, Japan, 2009
(and following pages)

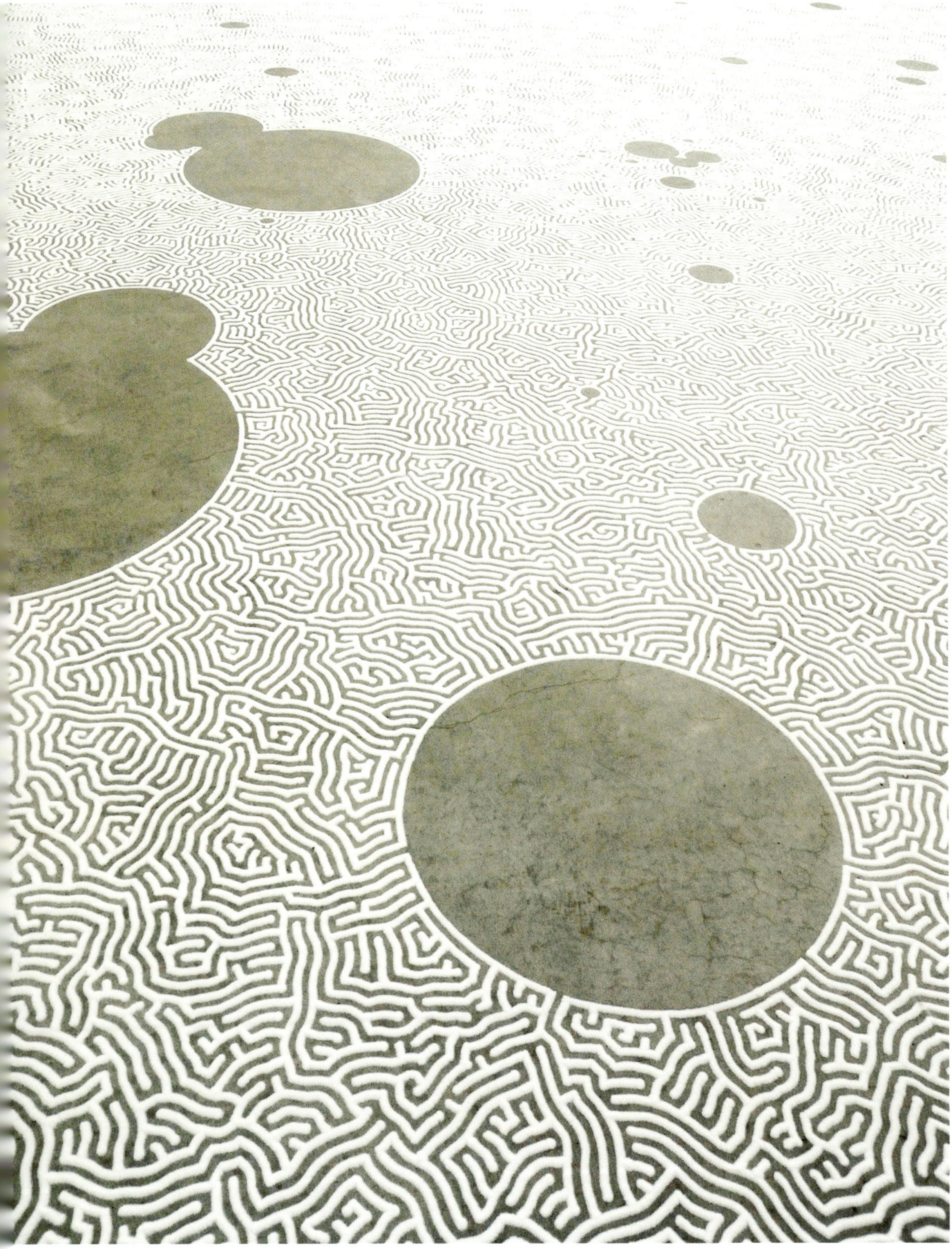

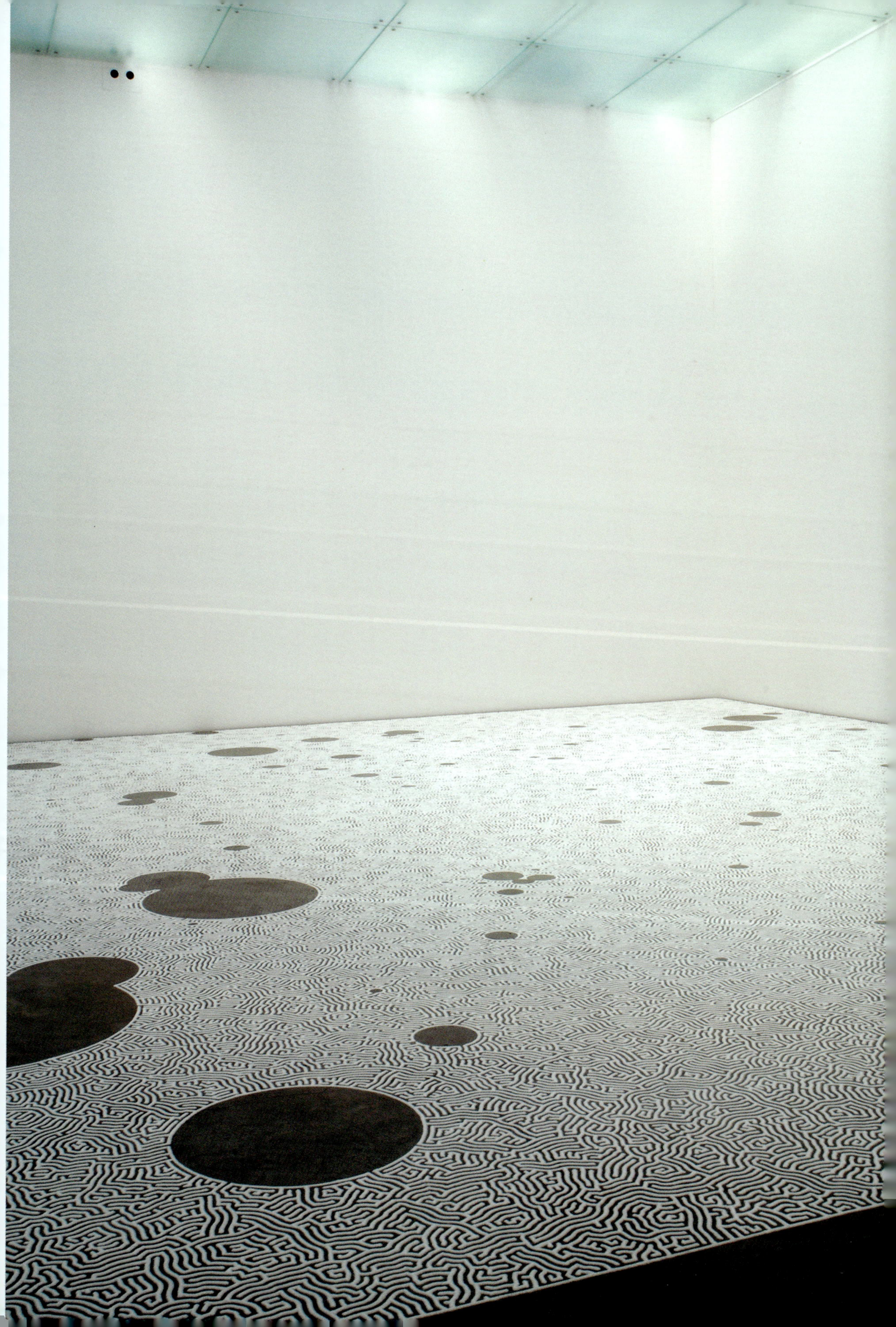

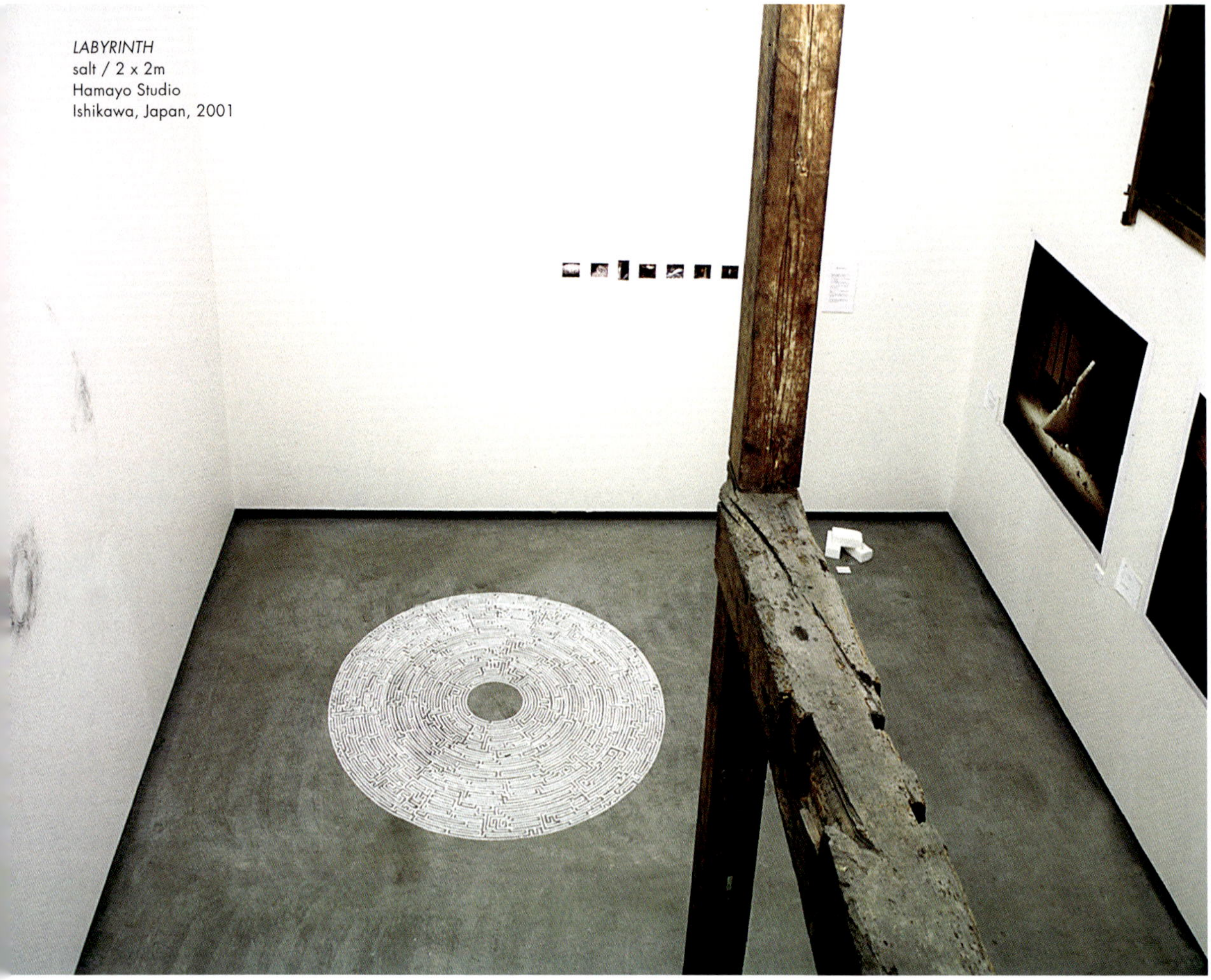

LABYRINTH
salt / 2 x 2m
Hamayo Studio
Ishikawa, Japan, 2001

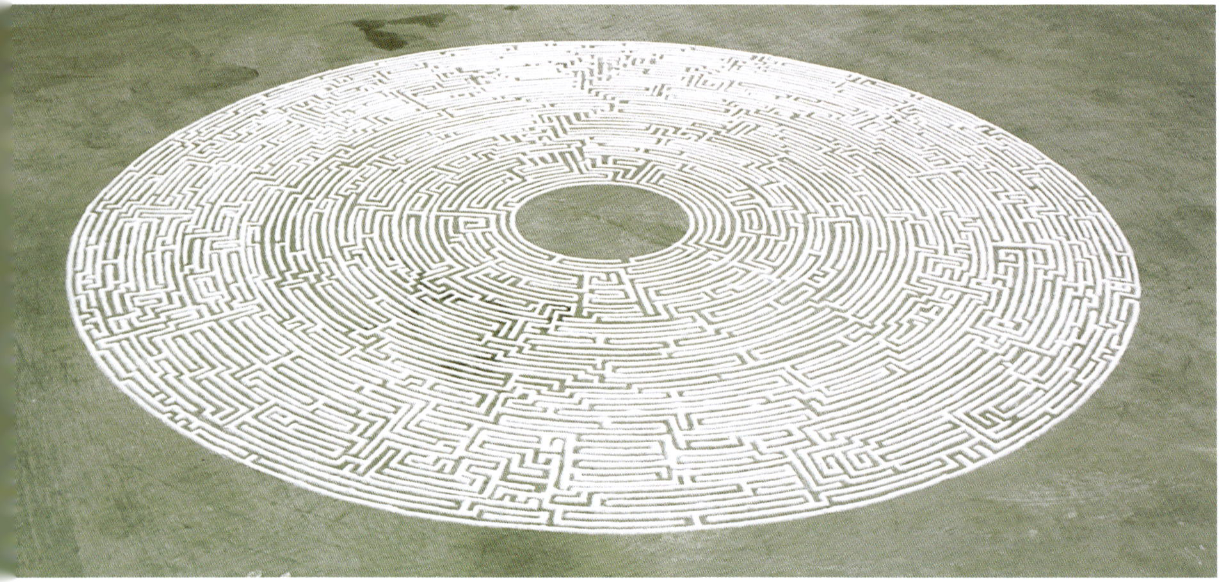

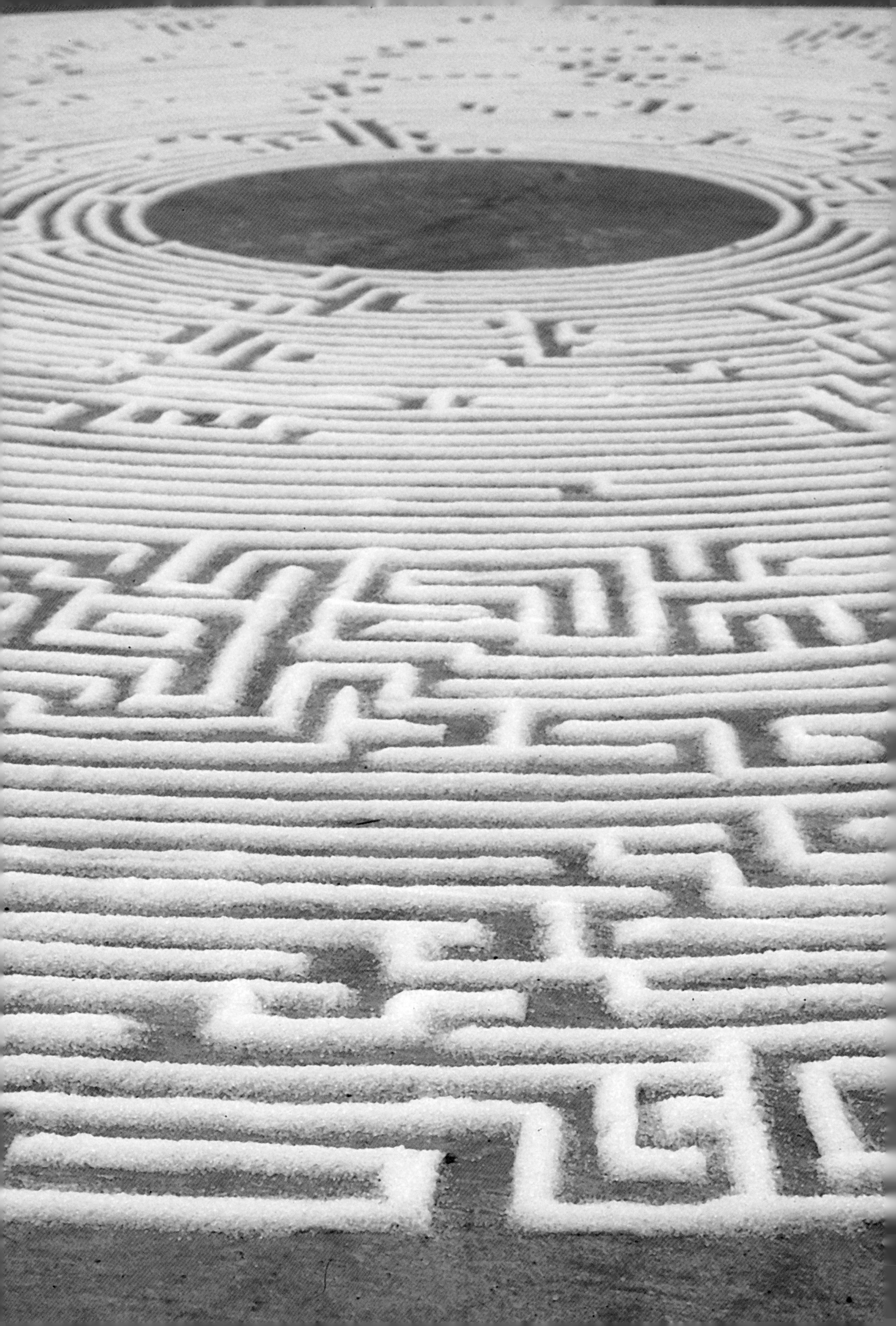

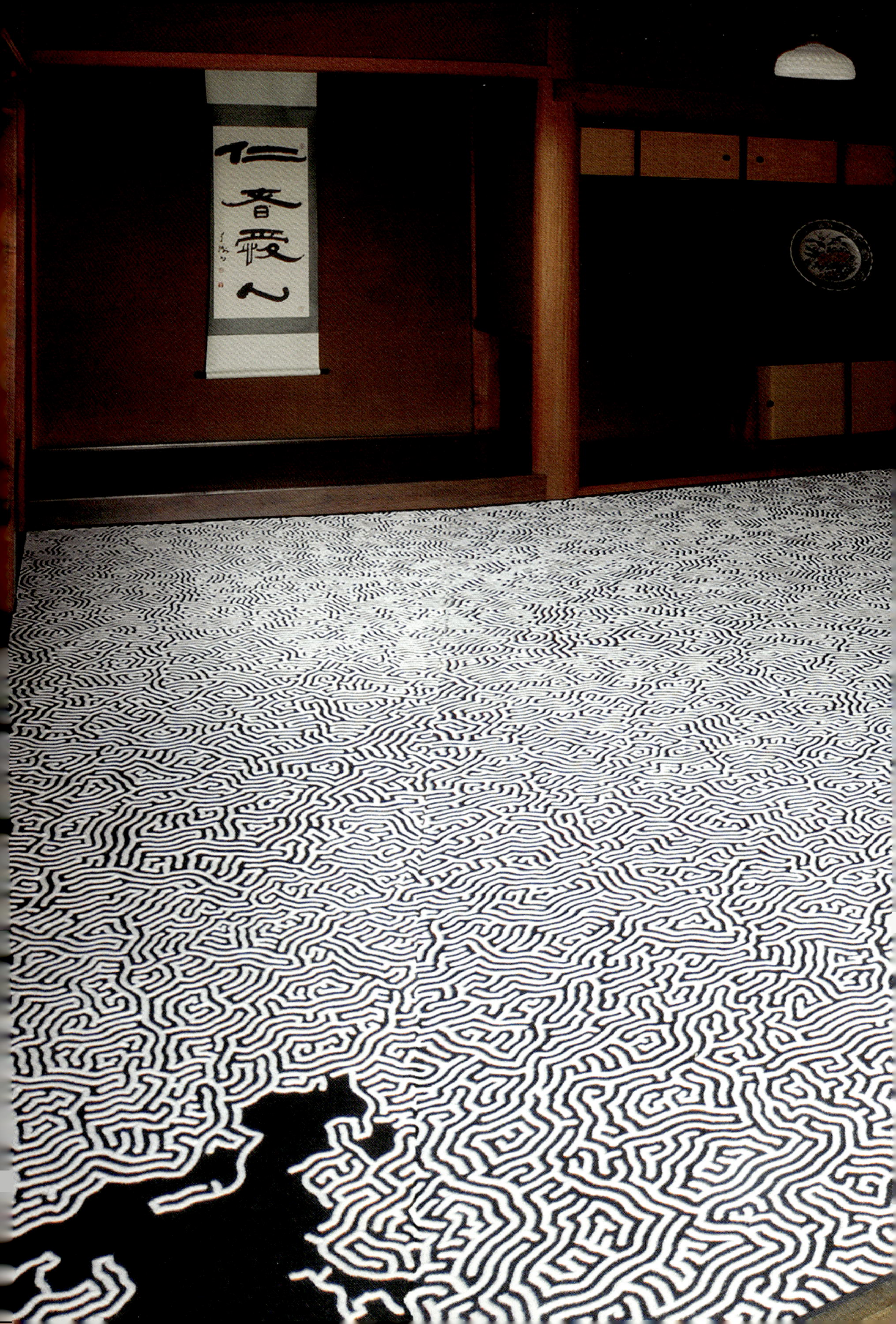
仁者愛人

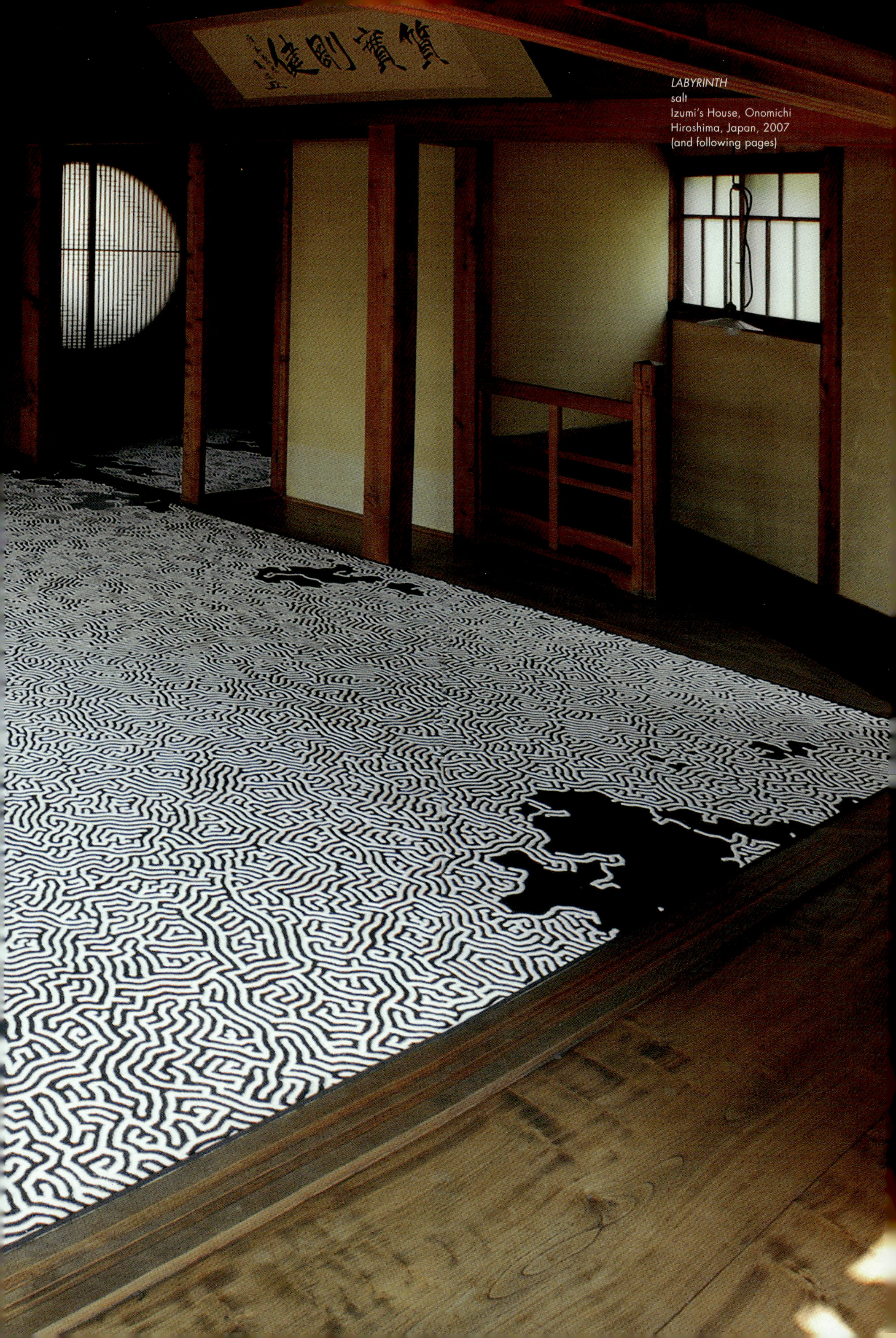

LABYRINTH
salt
Izumi's House, Onomichi
Hiroshima, Japan, 2007
(and following pages)

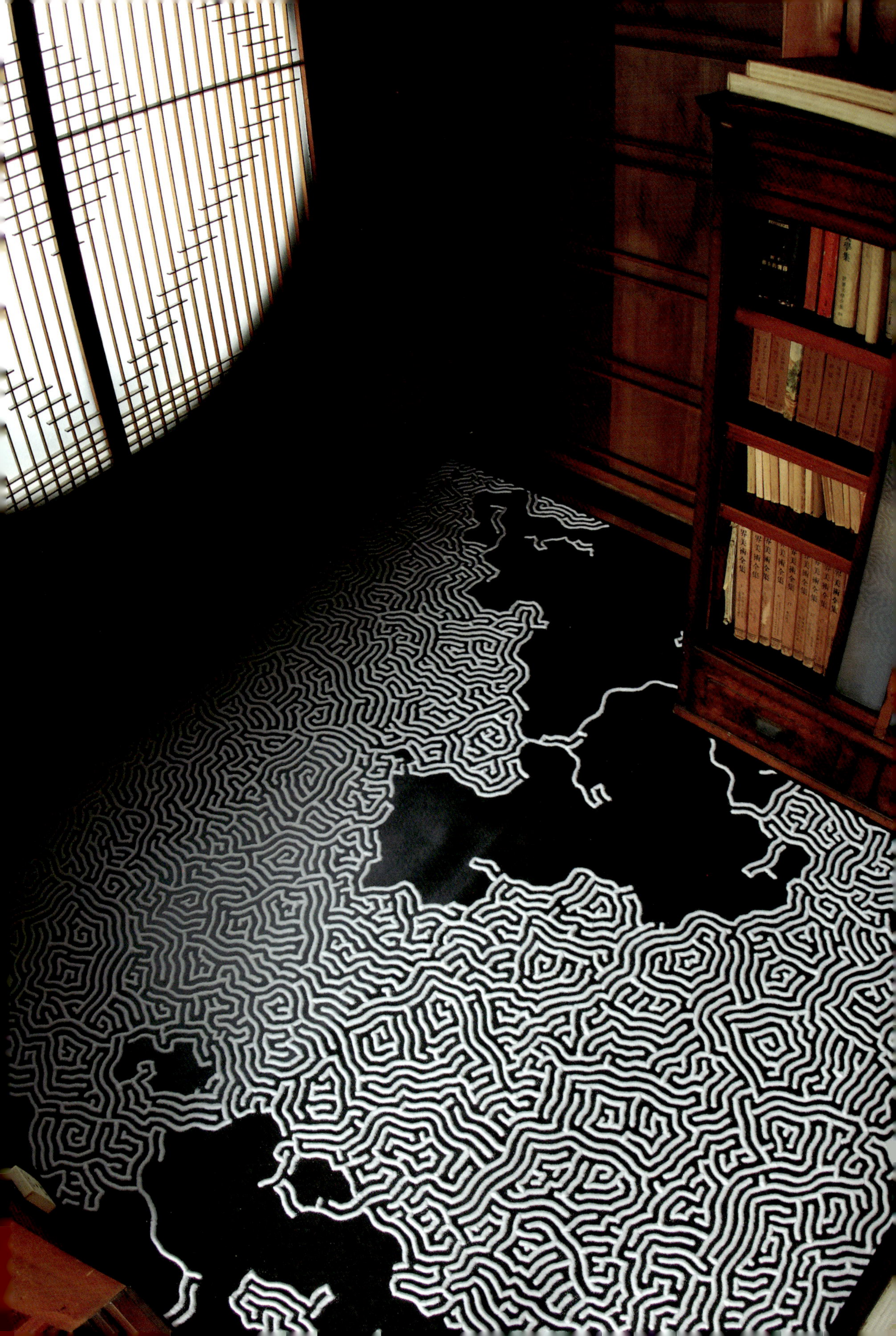

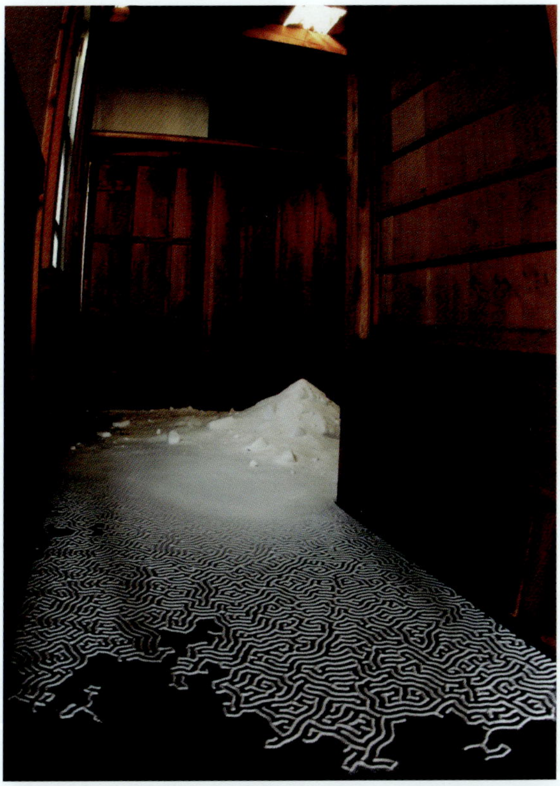
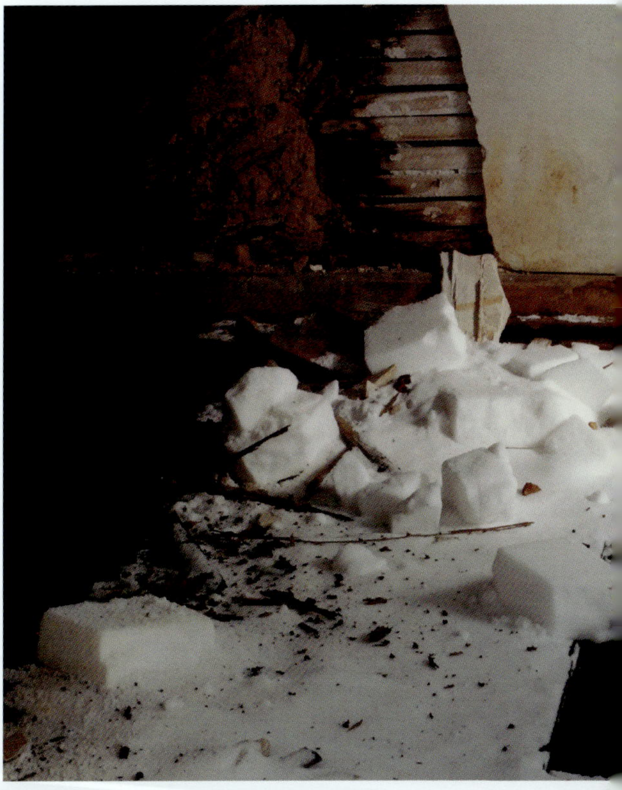
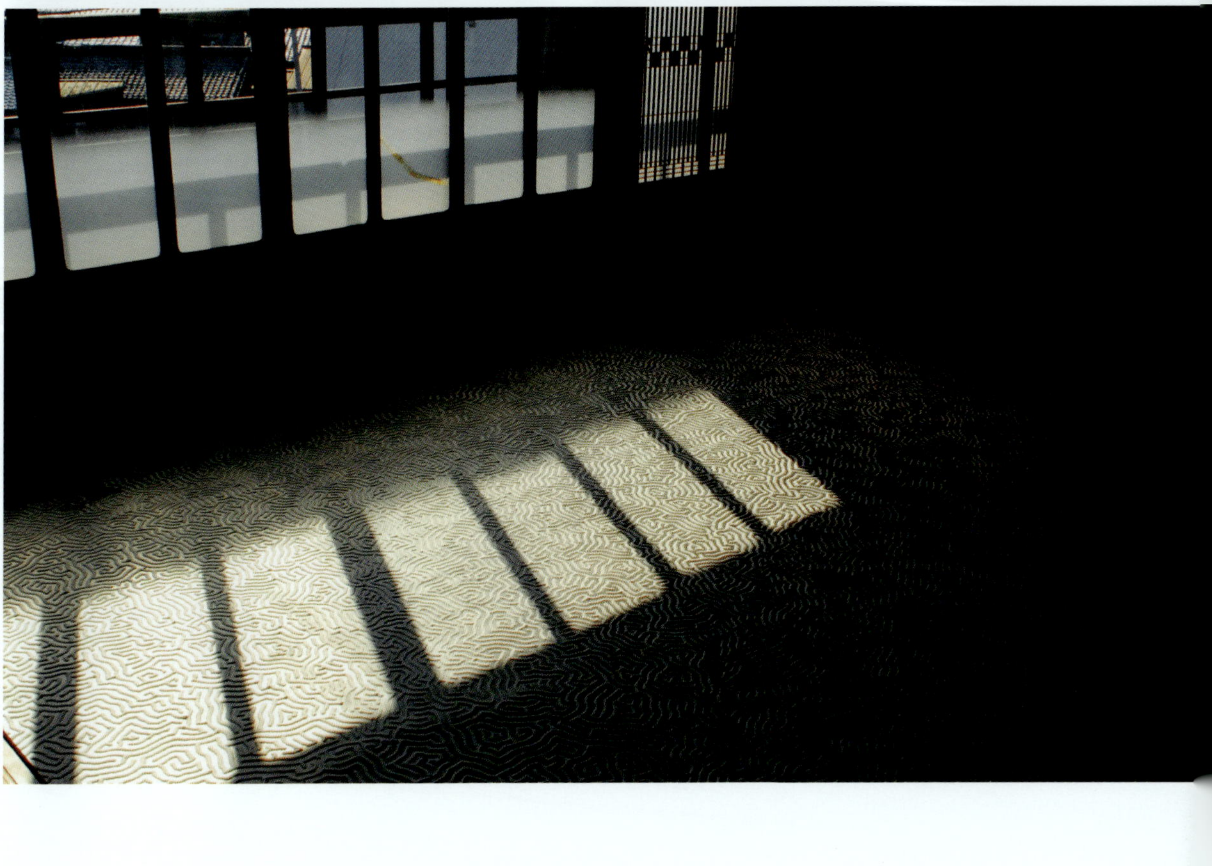

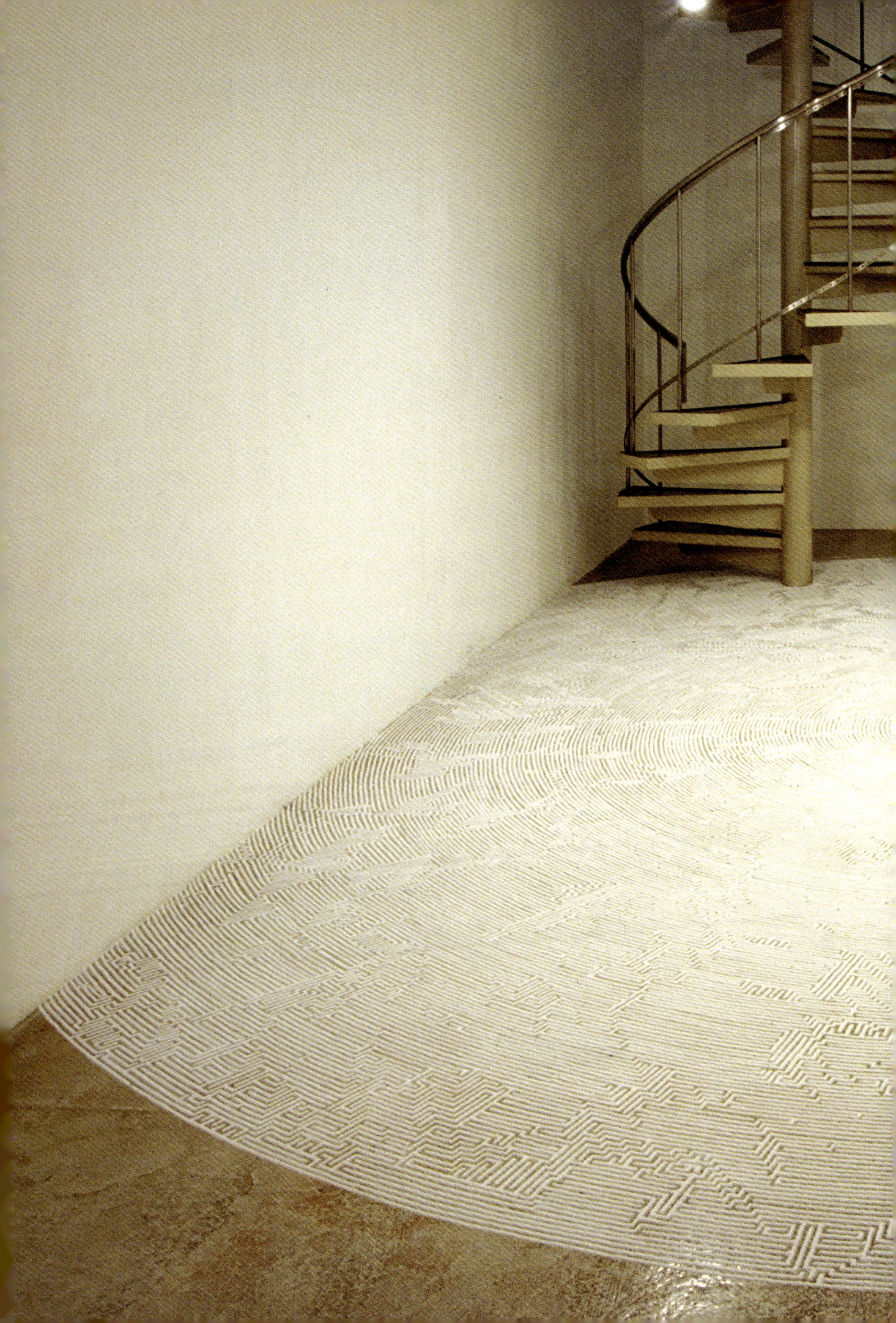

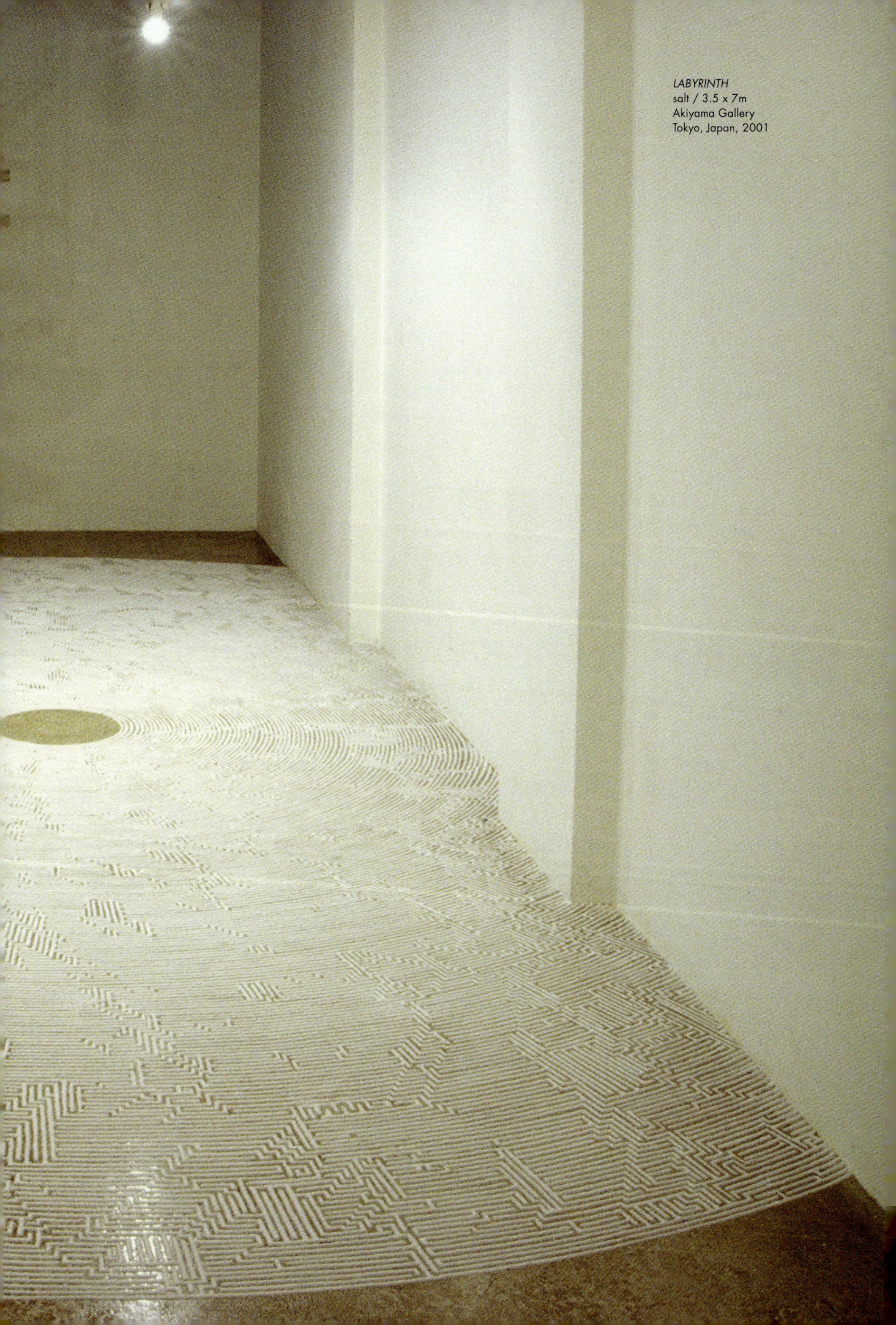

LABYRINTH
salt / 3.5 x 7m
Akiyama Gallery
Tokyo, Japan, 2001

LABYRINTH
salt / 17.3 x 13m
Museum of Contemporary Art, Tokyo
Tokyo, Japan, 2010
(and following pages)

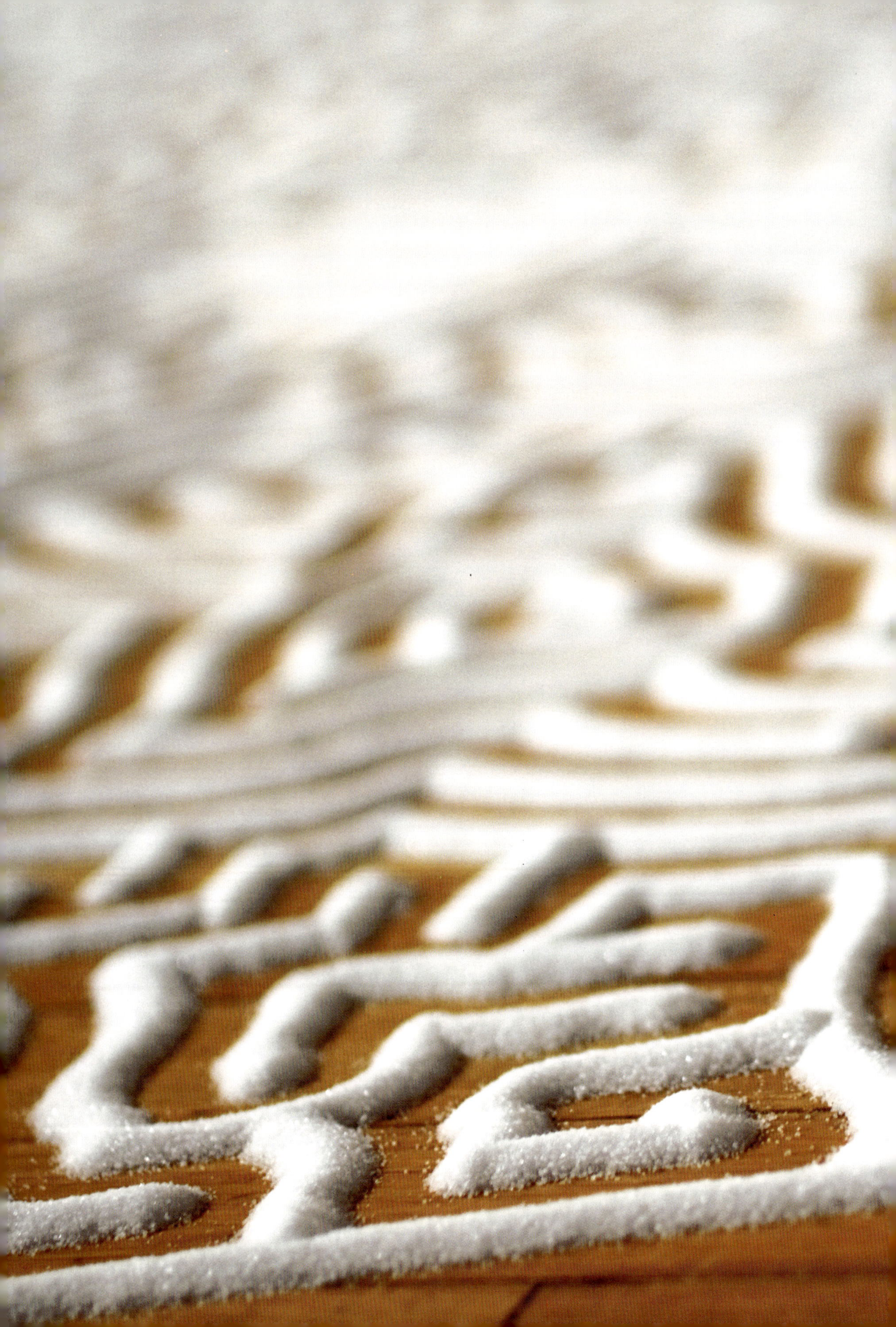

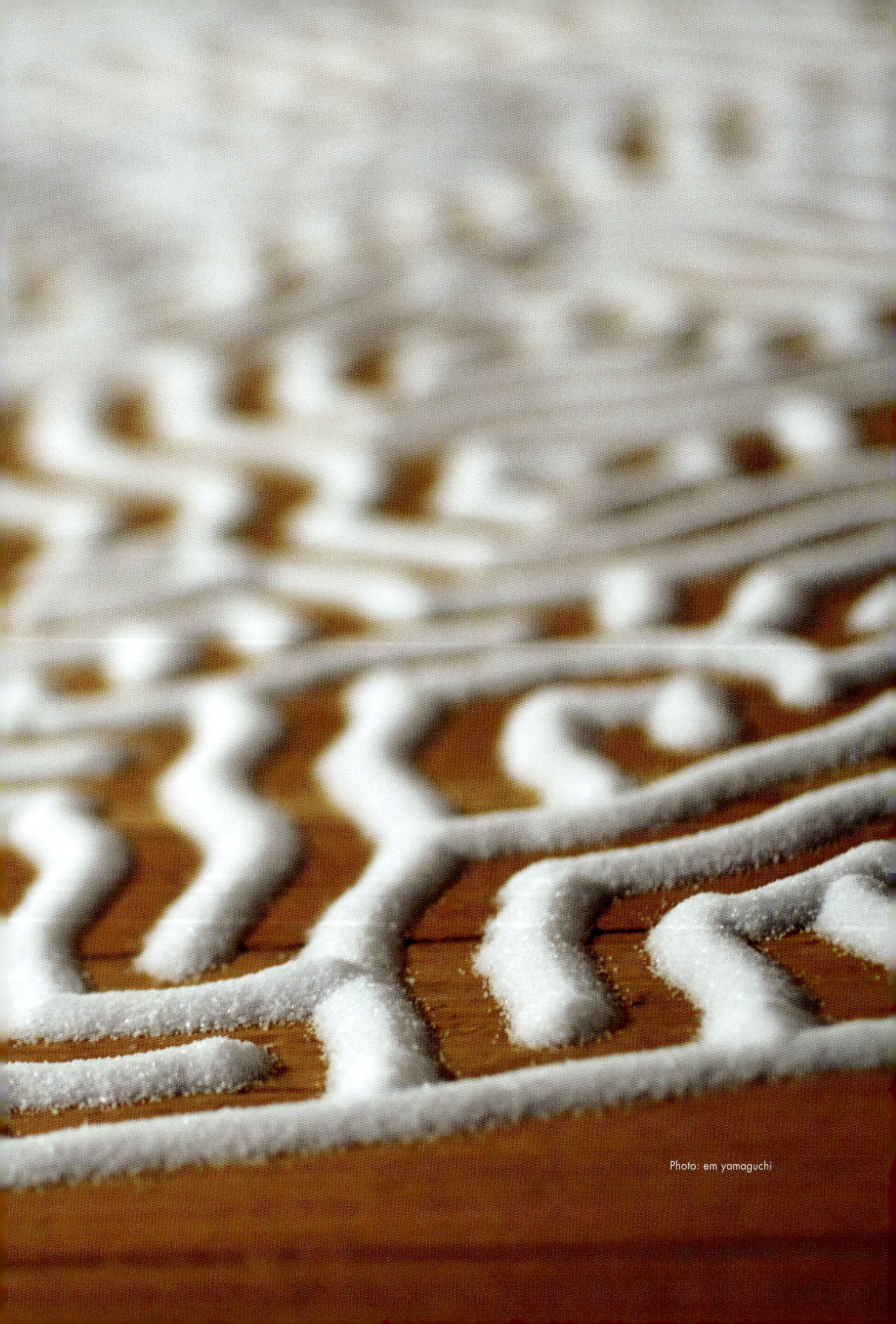

Photo: em yamaguchi

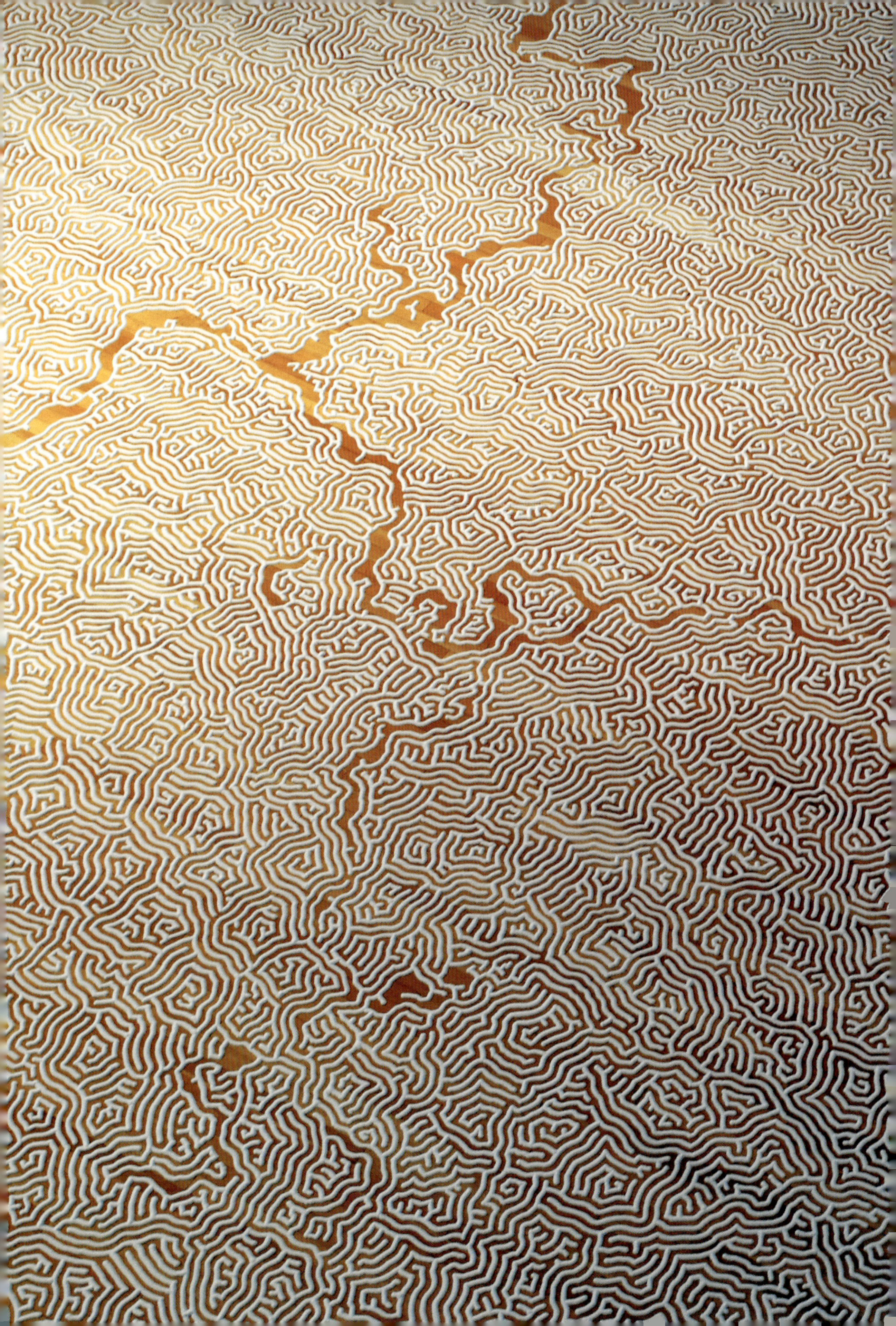

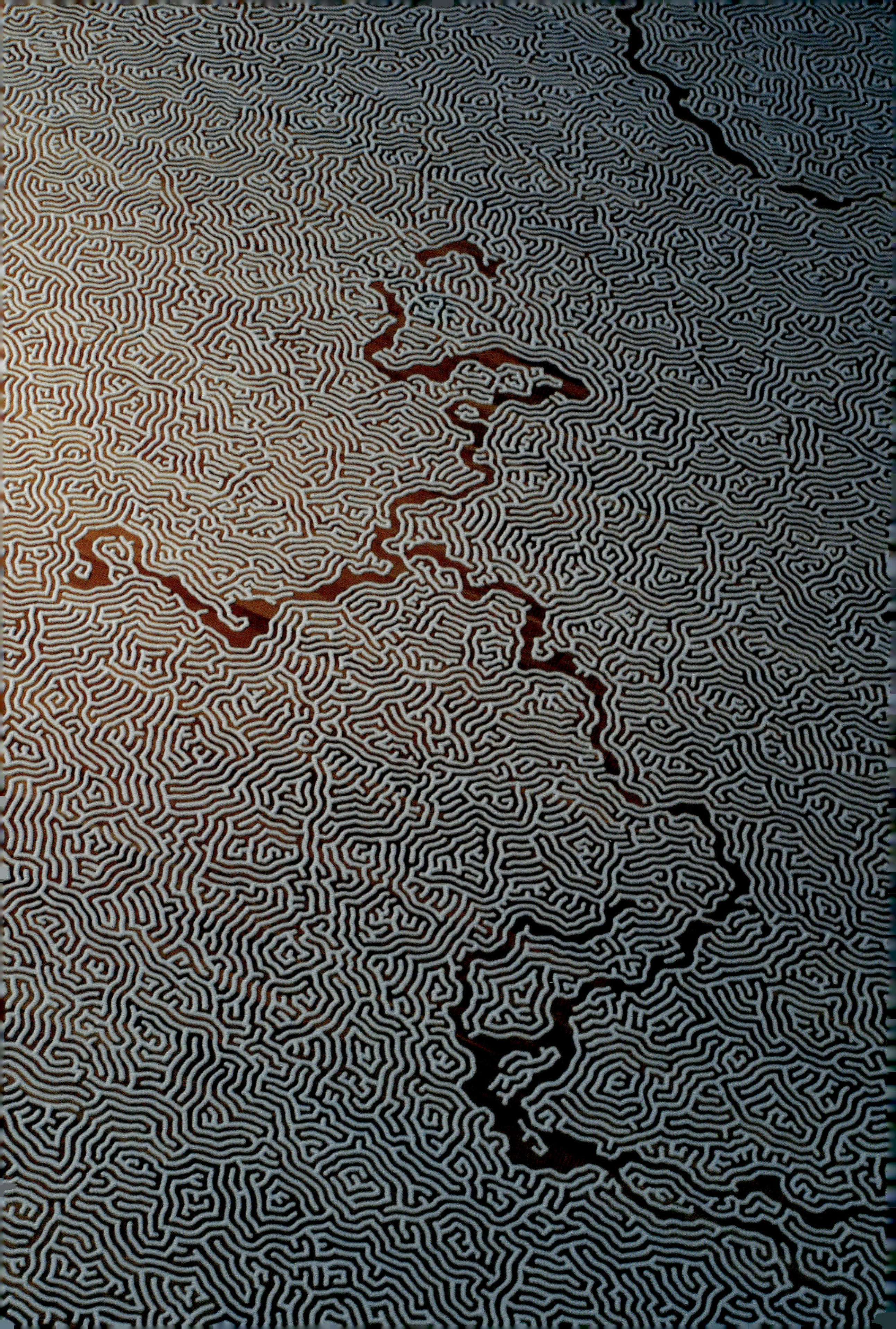

Garden

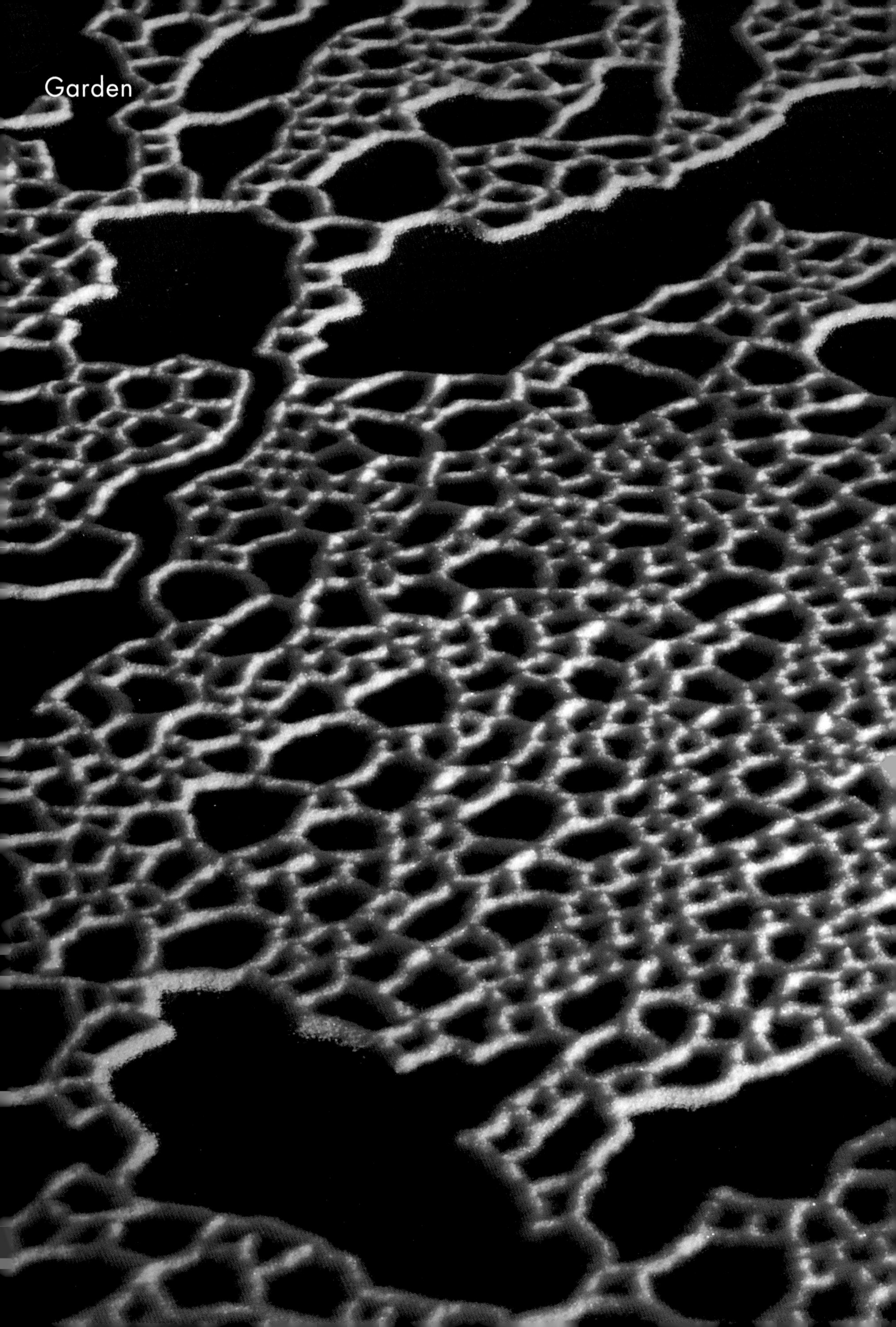

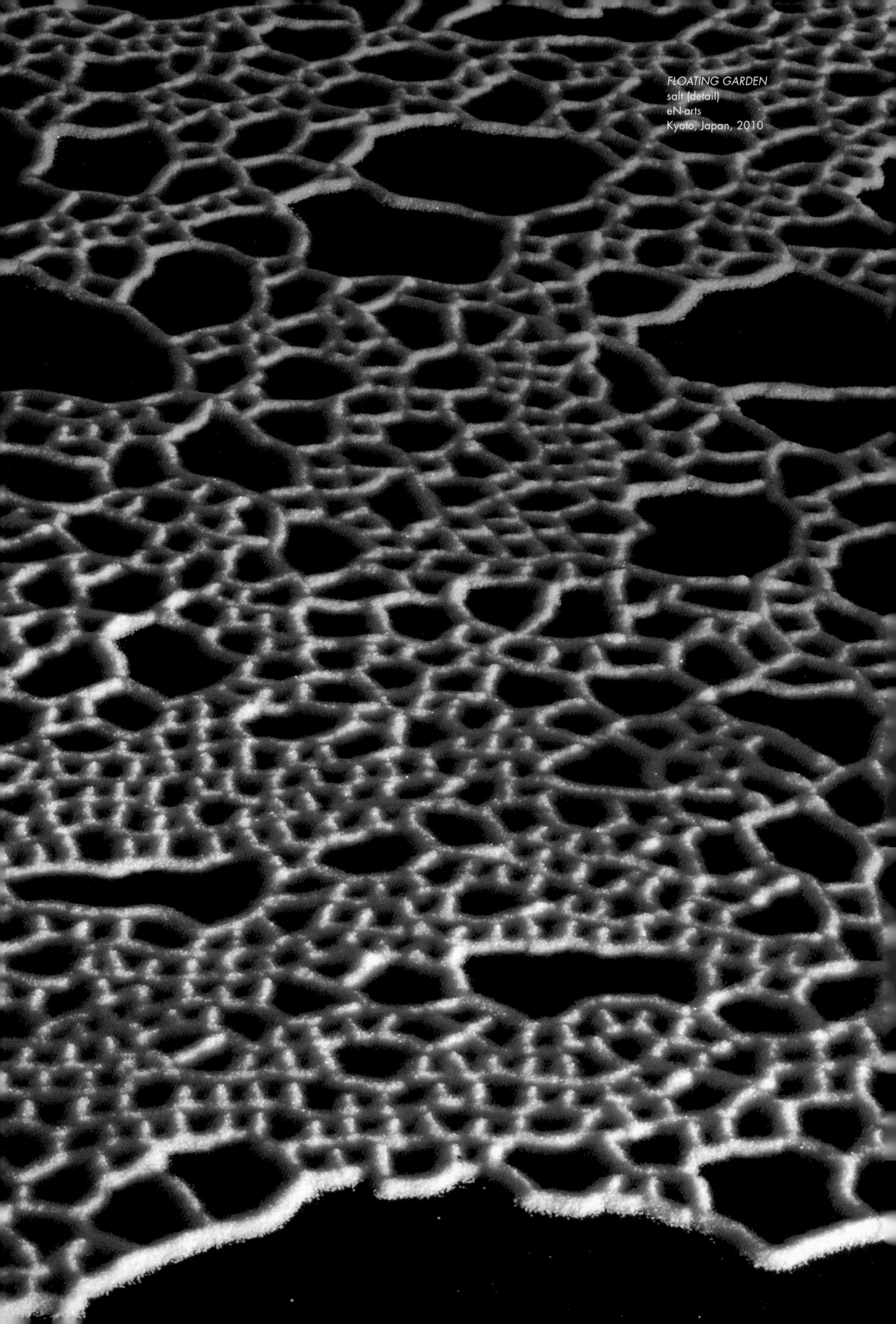

FLOATING GARDEN
salt (detail)
eN-arts
Kyoto, Japan, 2010

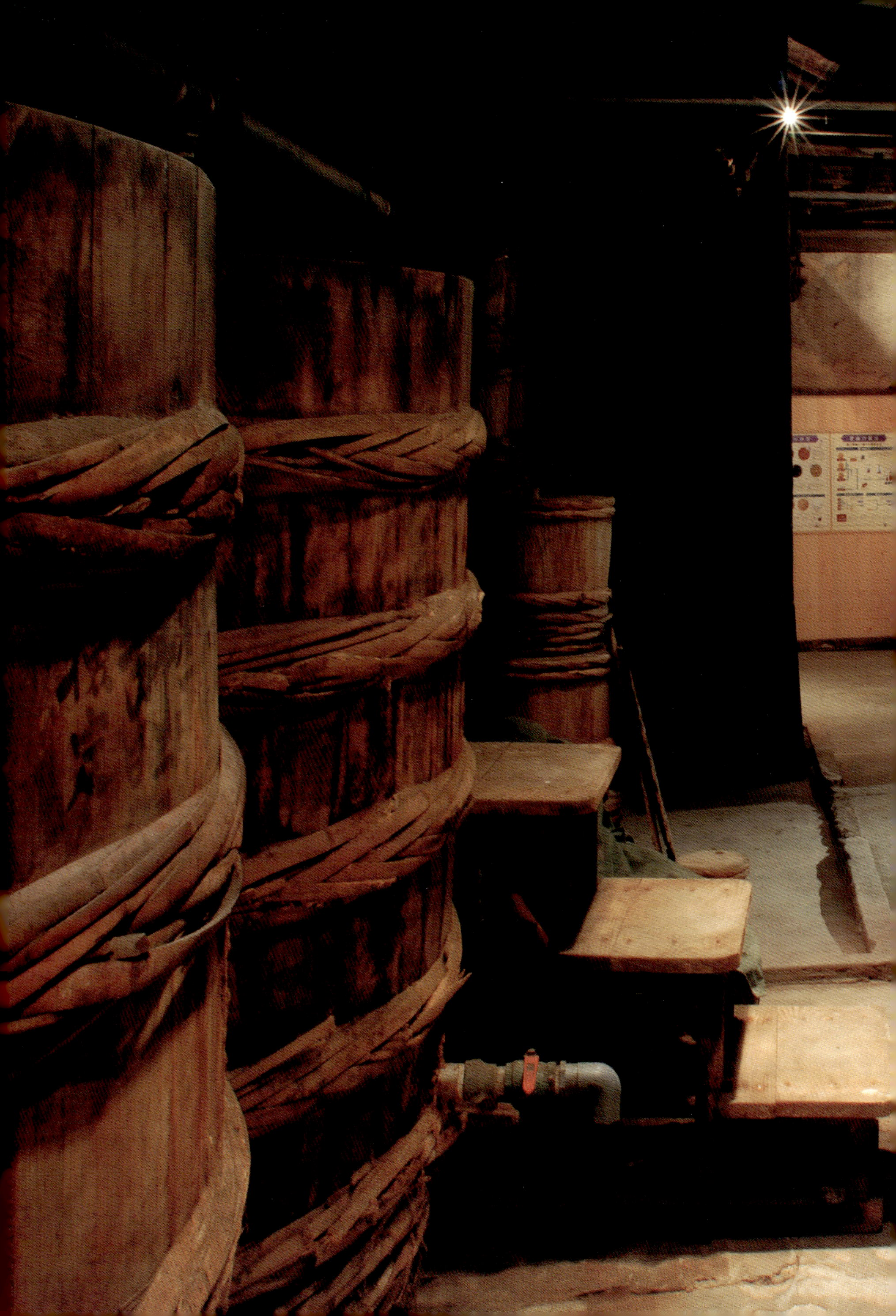

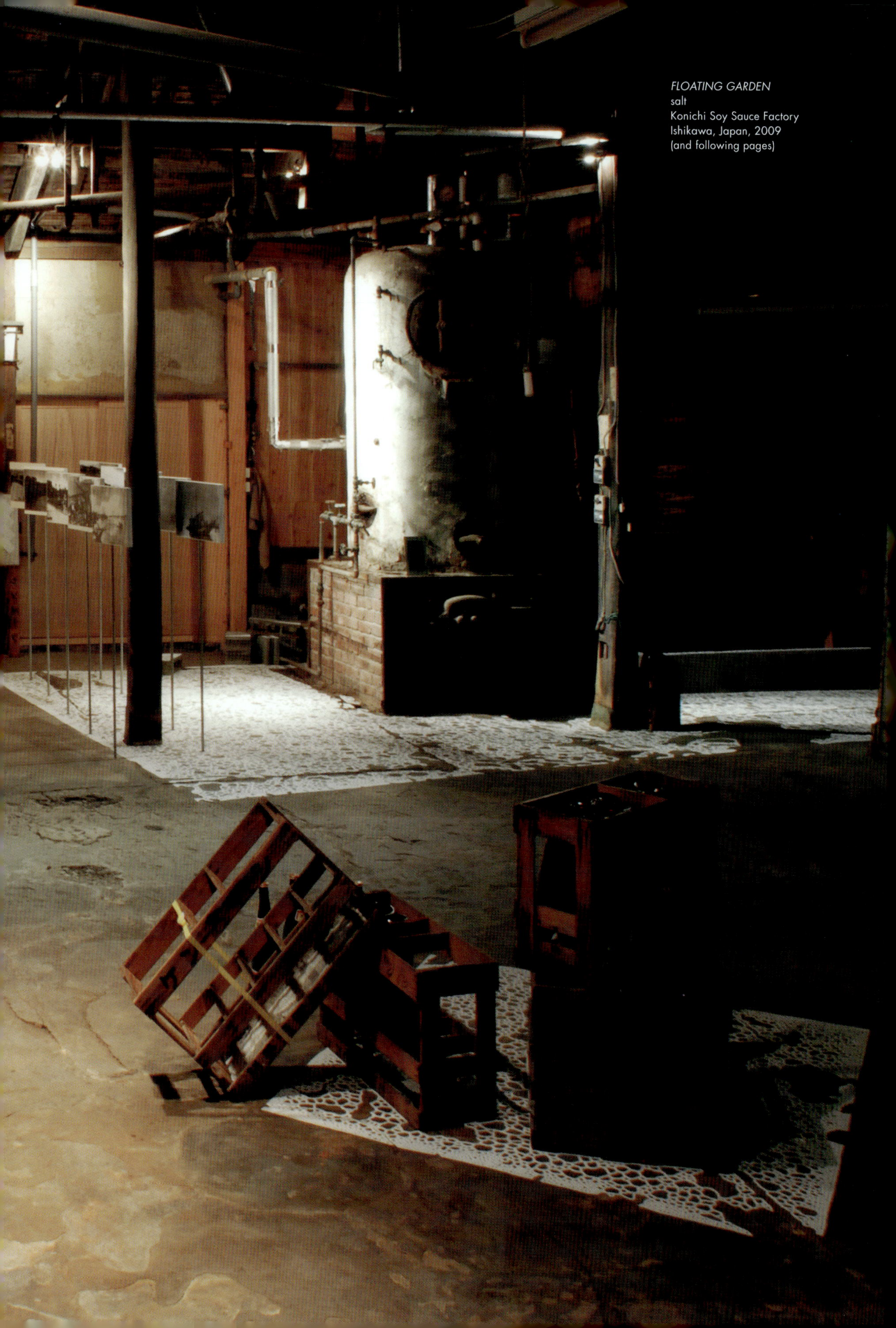

FLOATING GARDEN
salt
Konichi Soy Sauce Factory
Ishikawa, Japan, 2009
(and following pages)

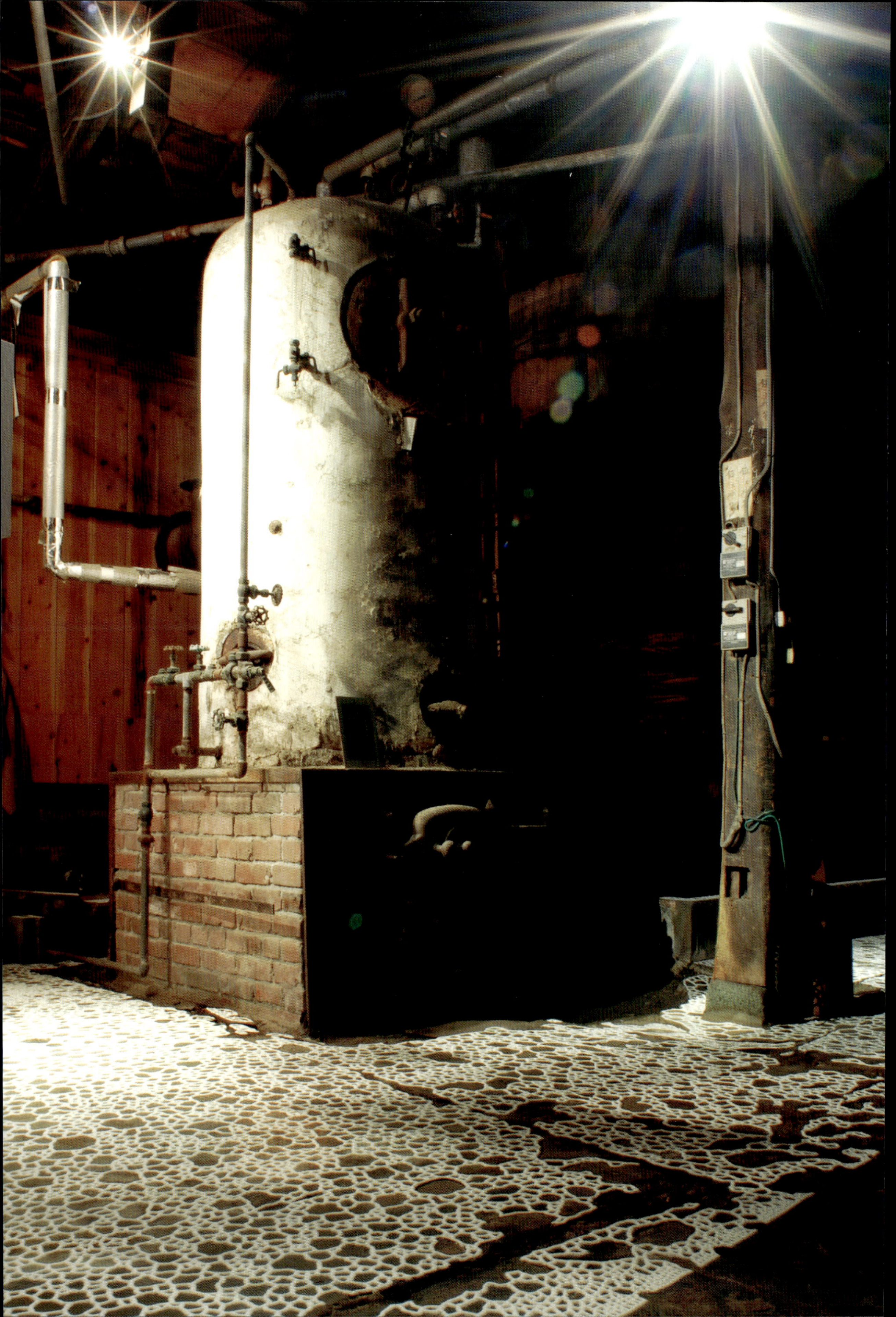

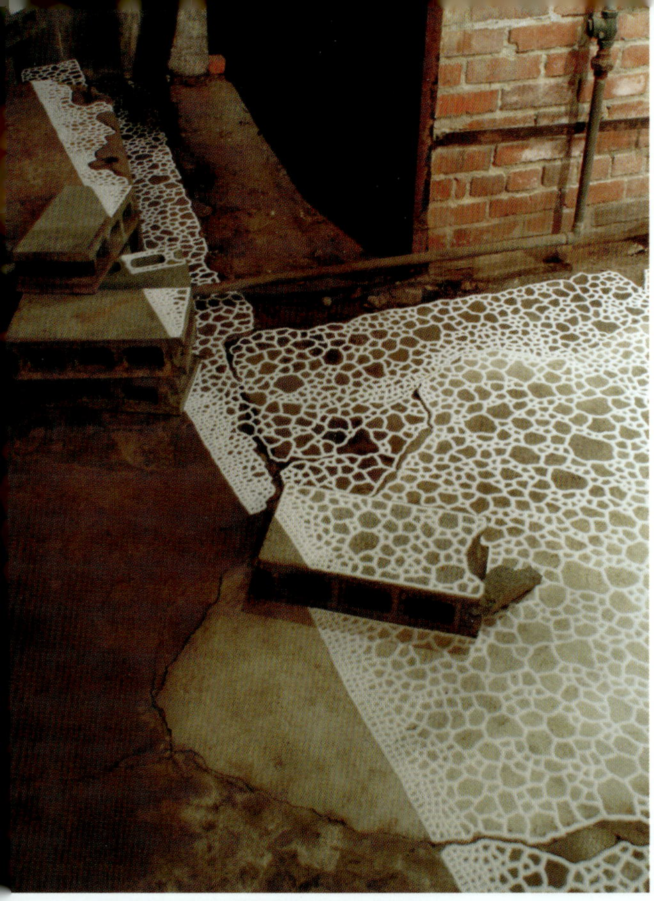

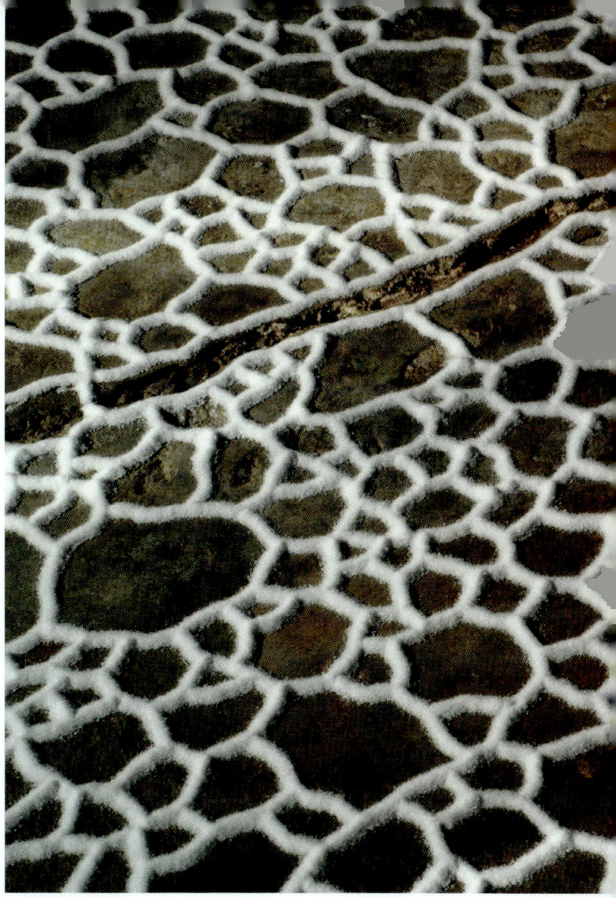

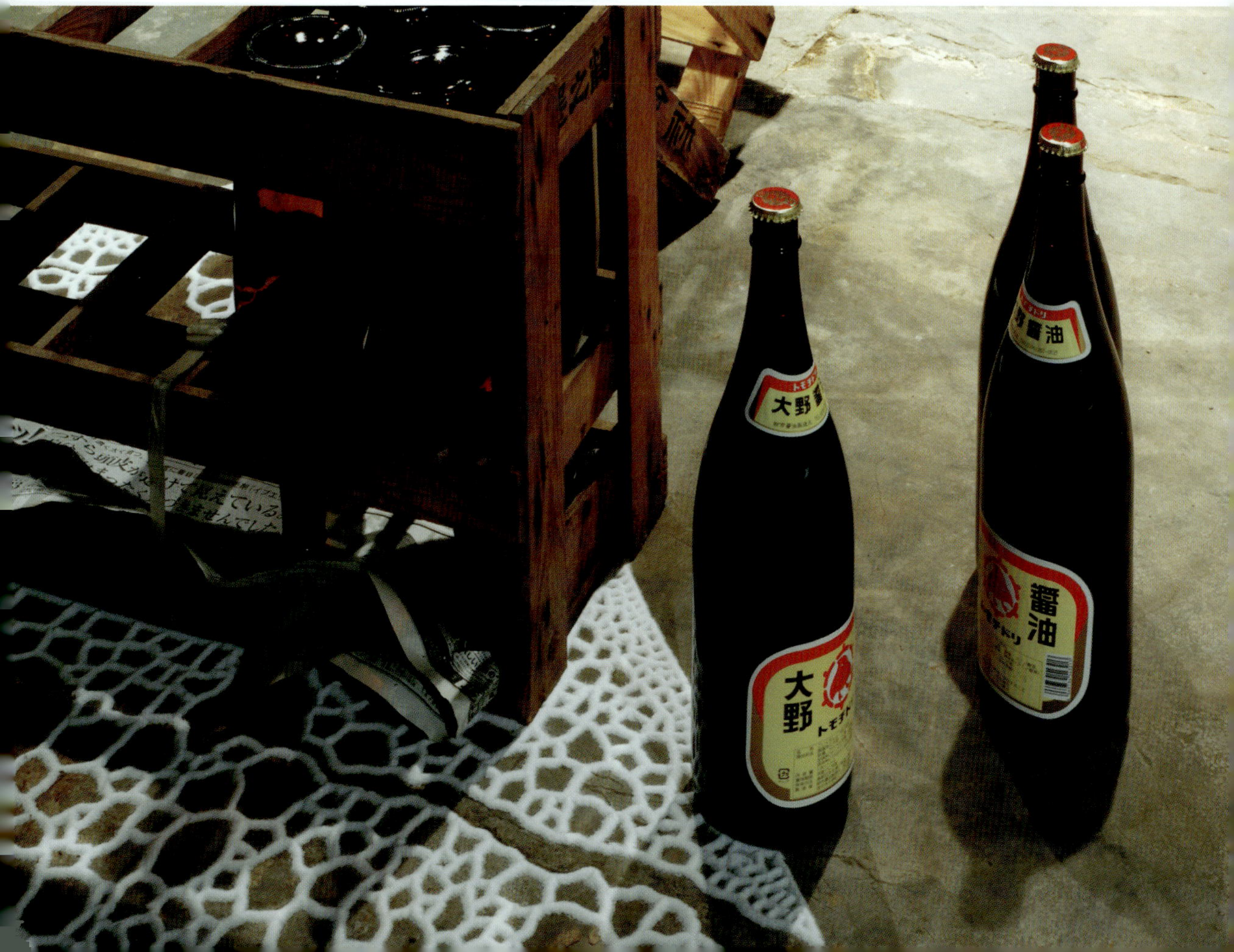
大野
醤油

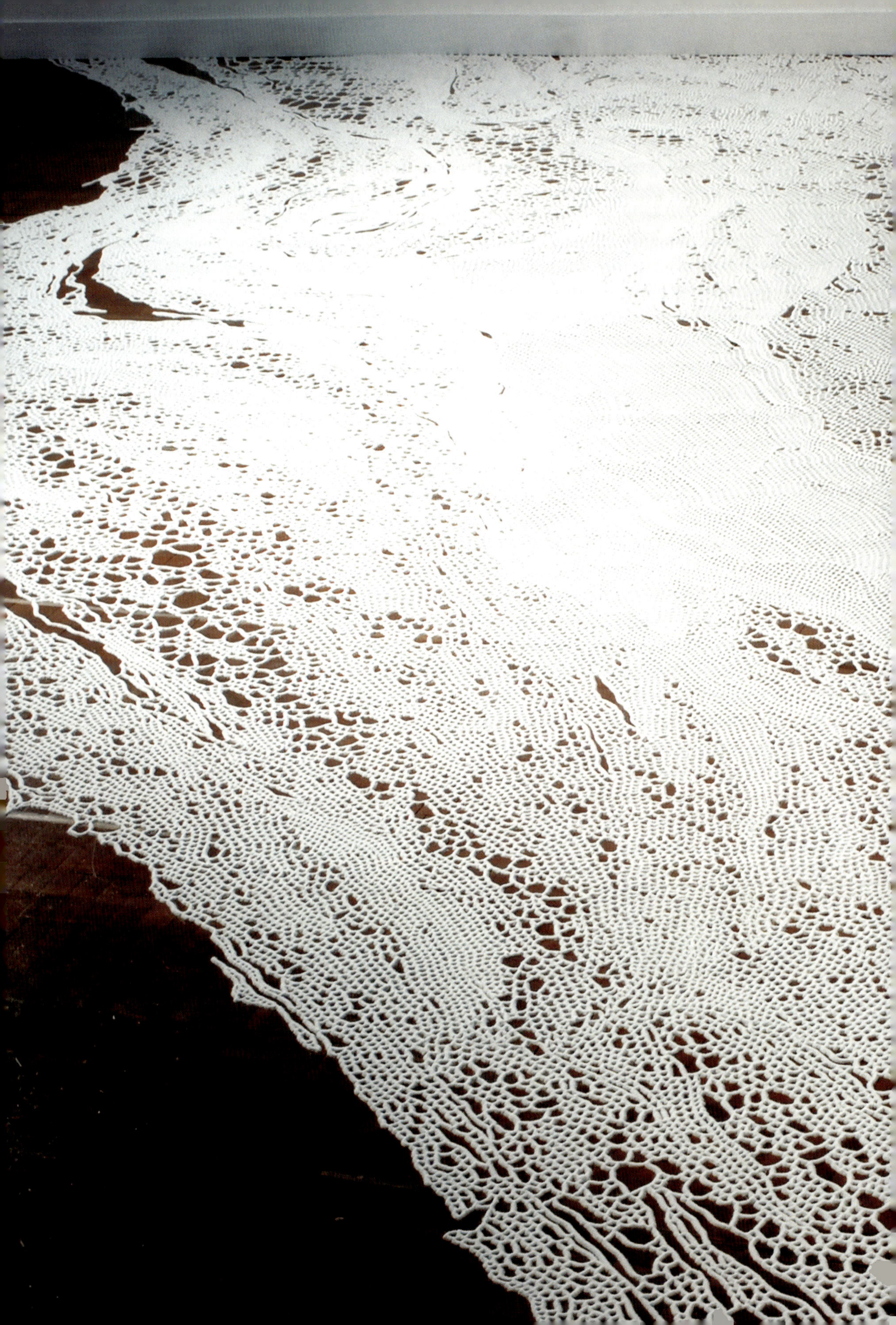

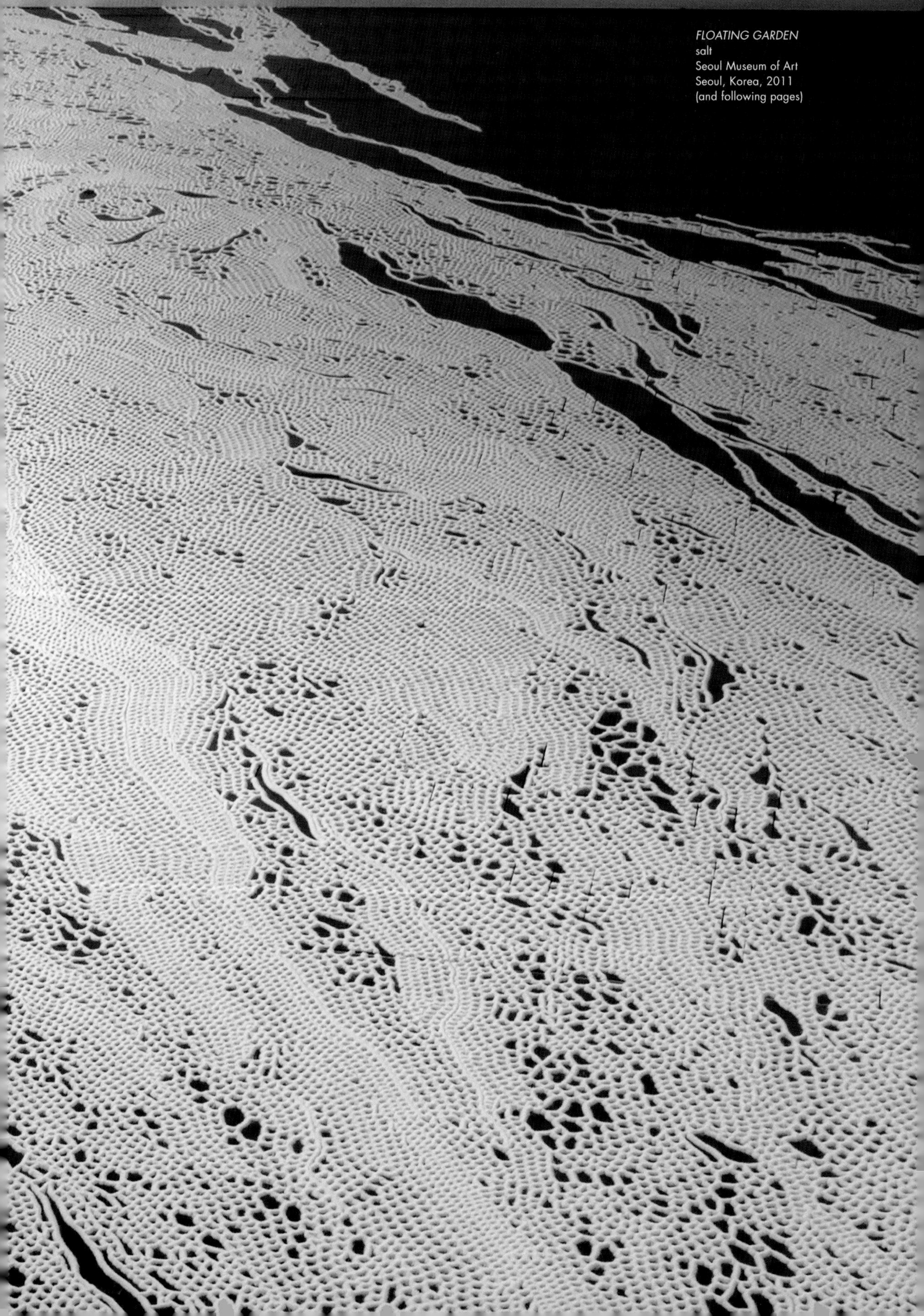

FLOATING GARDEN
salt
Seoul Museum of Art
Seoul, Korea, 2011
(and following pages)

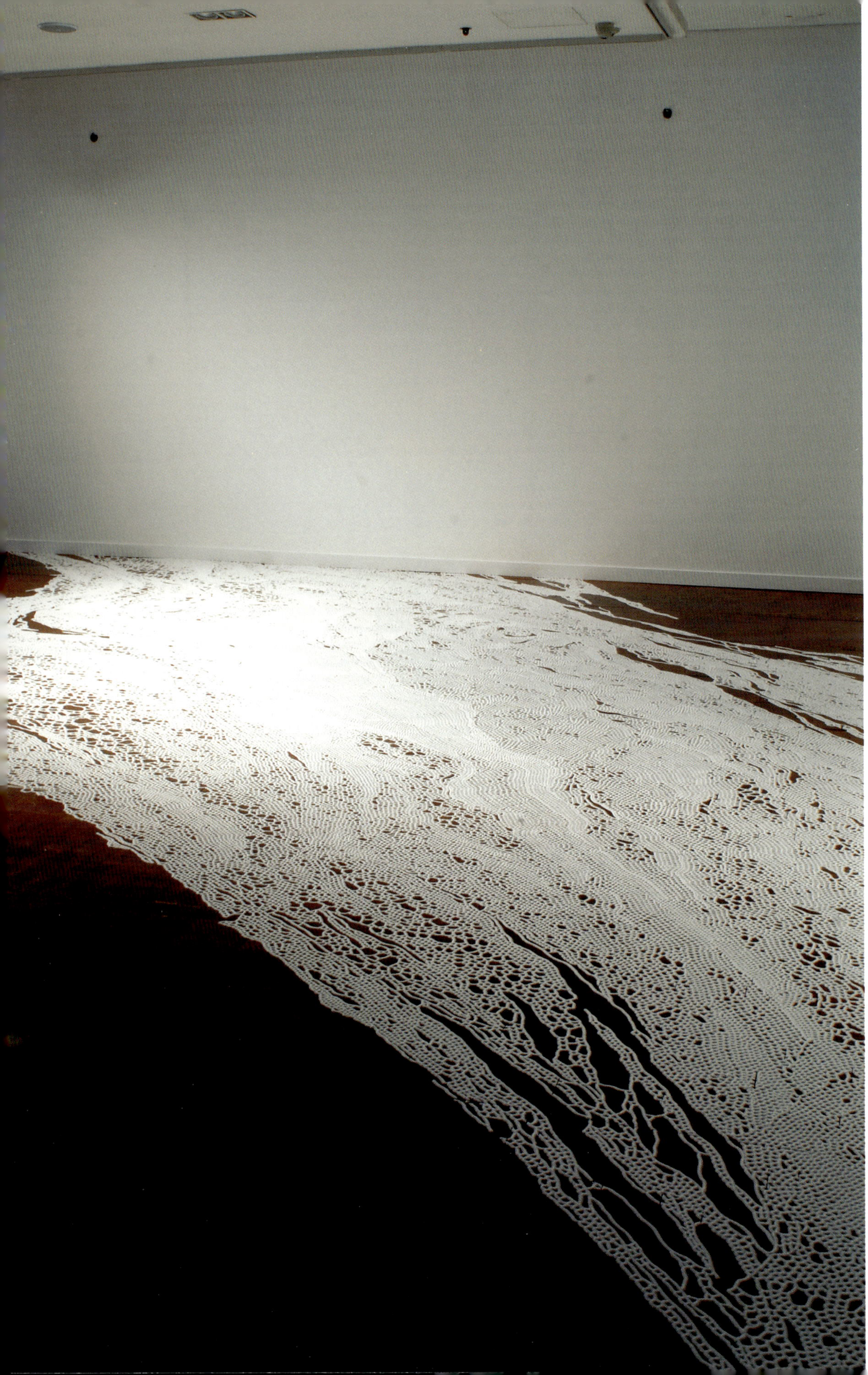

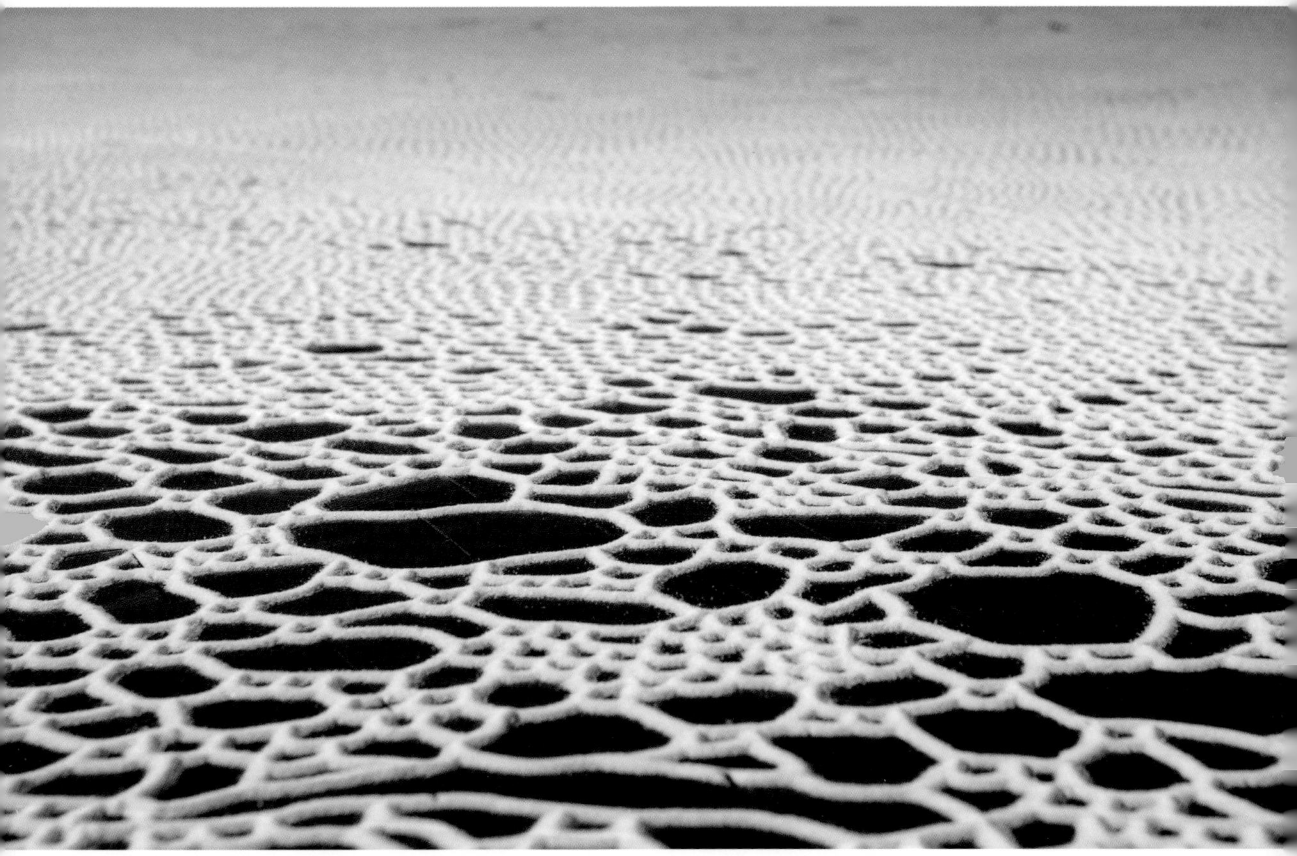

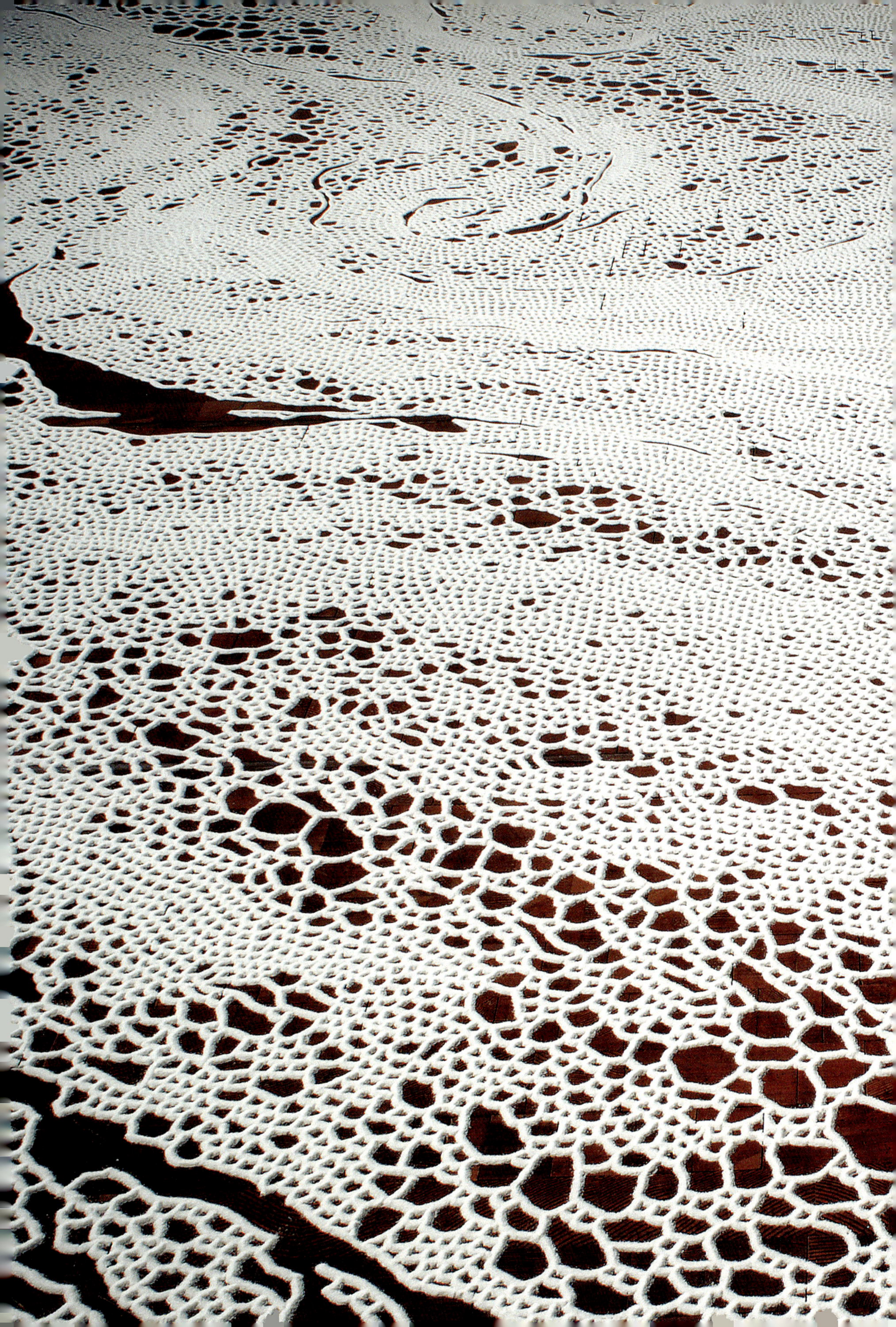

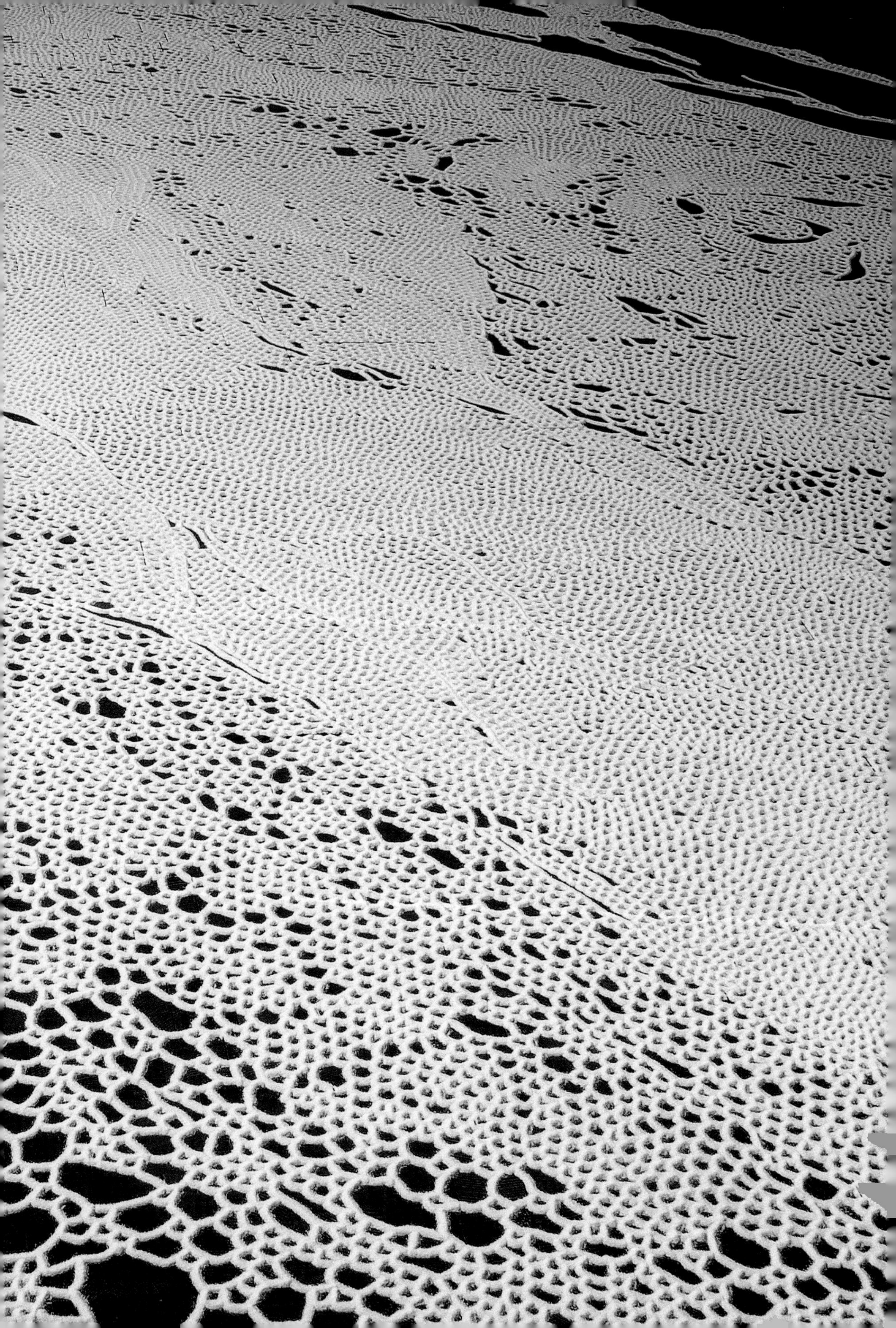

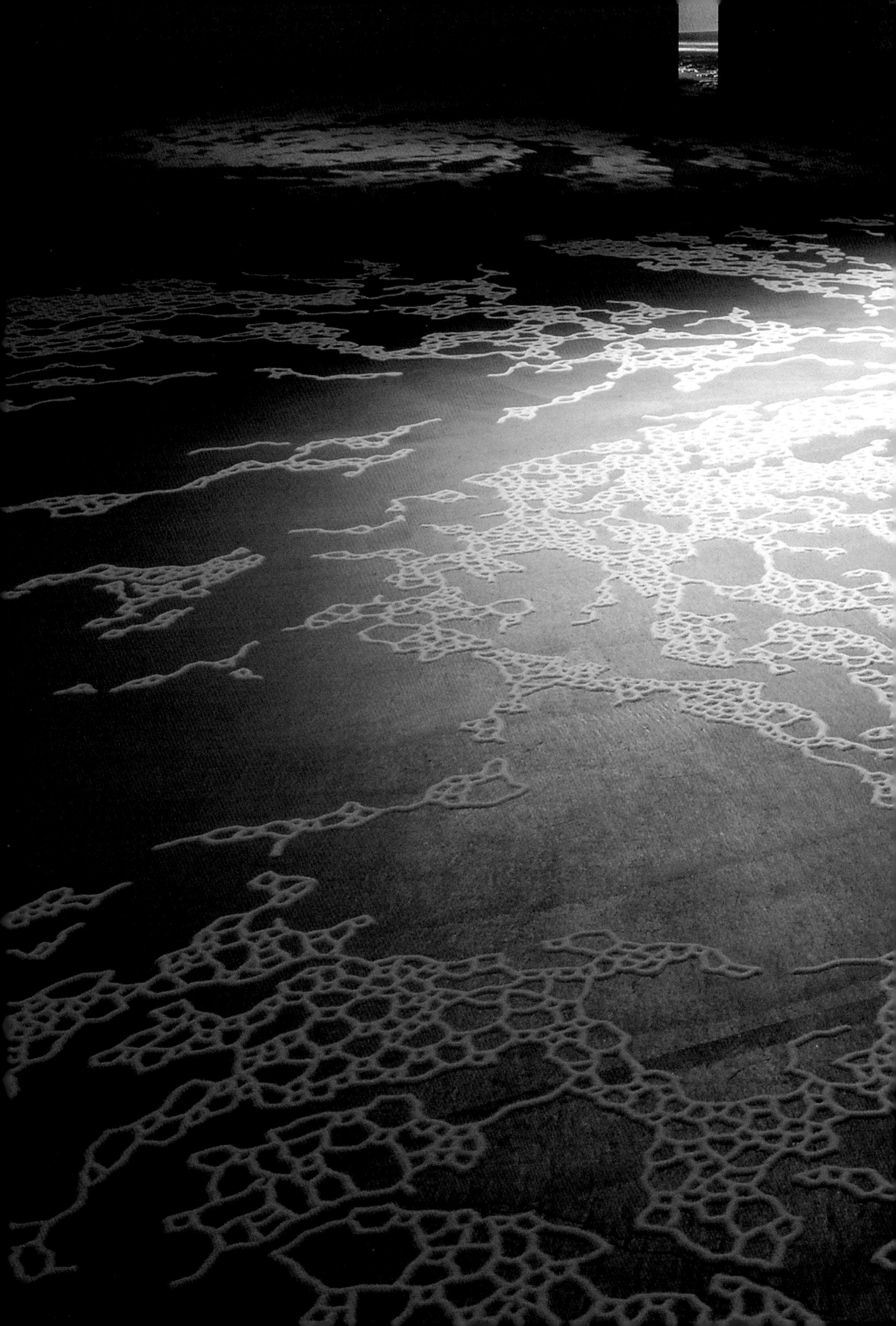

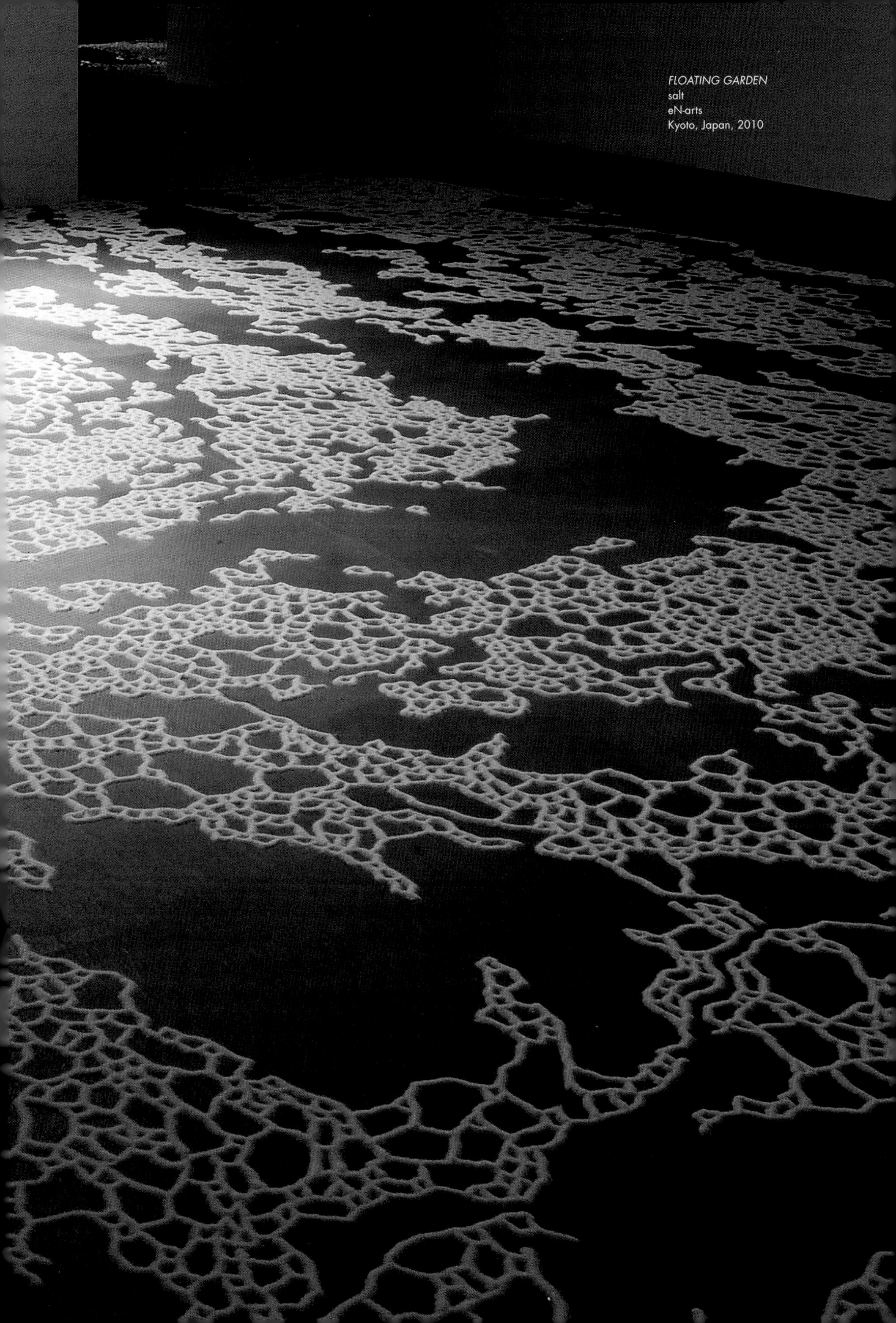

FLOATING GARDEN
salt
eN-arts
Kyoto, Japan, 2010

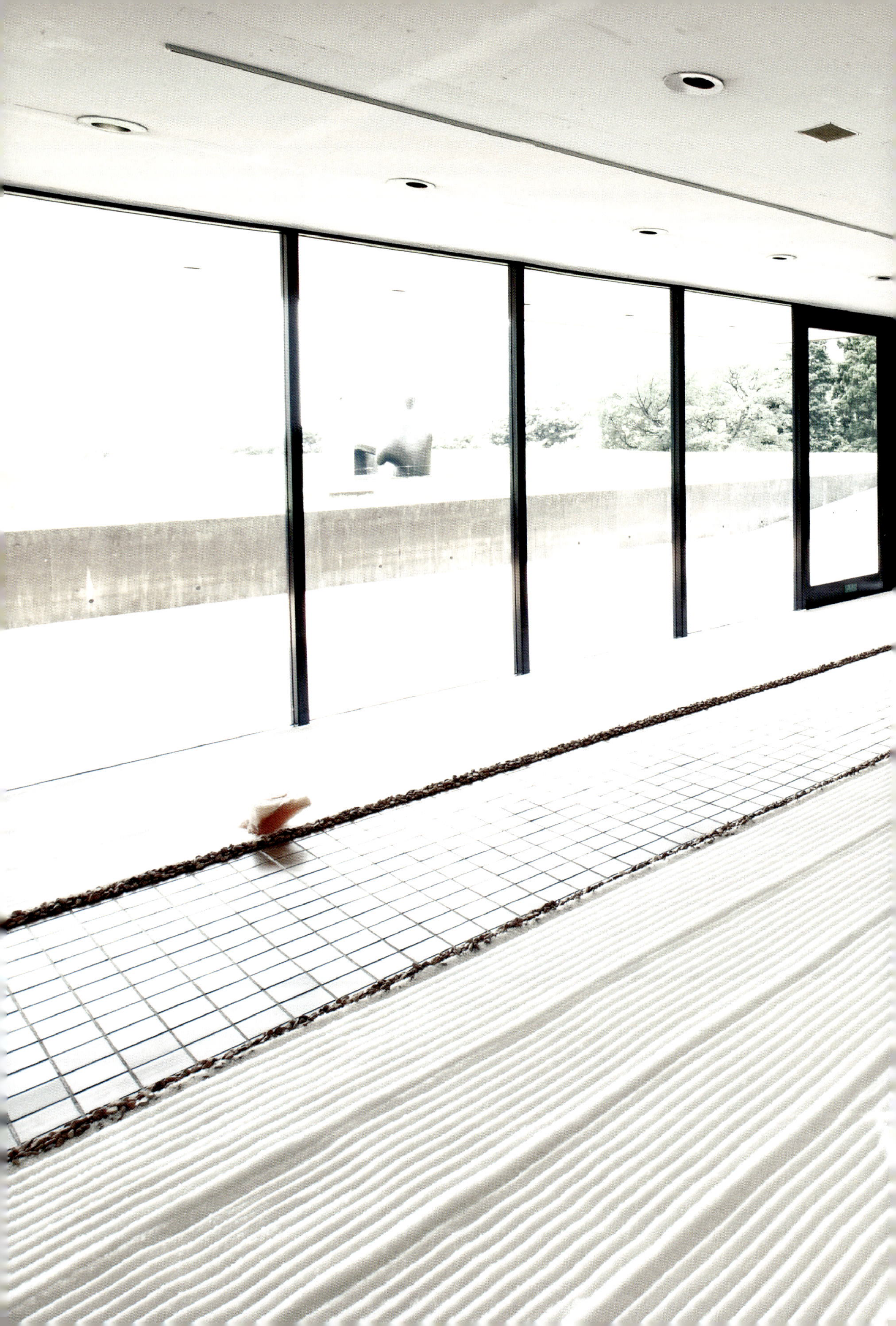

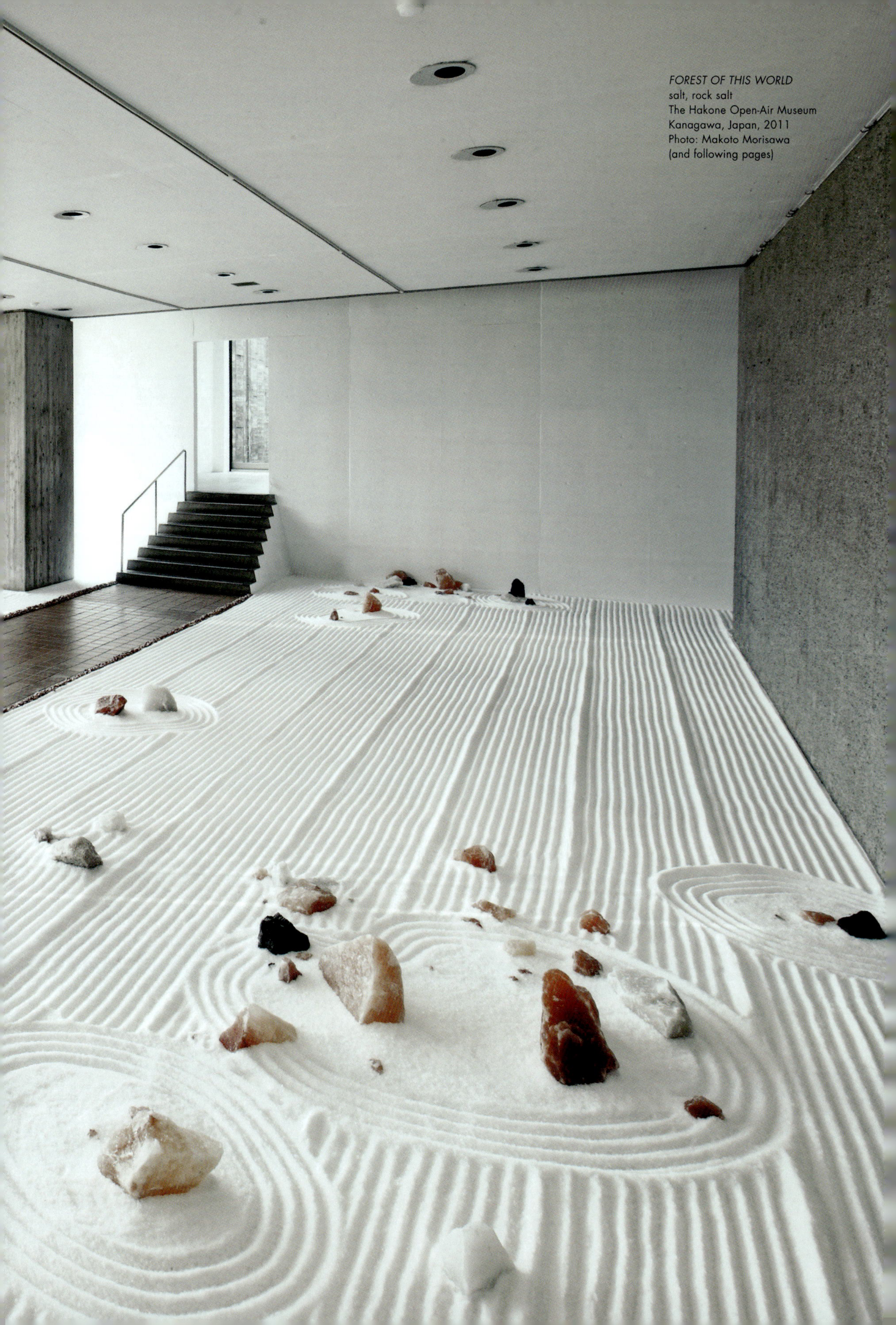

FOREST OF THIS WORLD
salt, rock salt
The Hakone Open-Air Museum
Kanagawa, Japan, 2011
Photo: Makoto Morisawa
(and following pages)

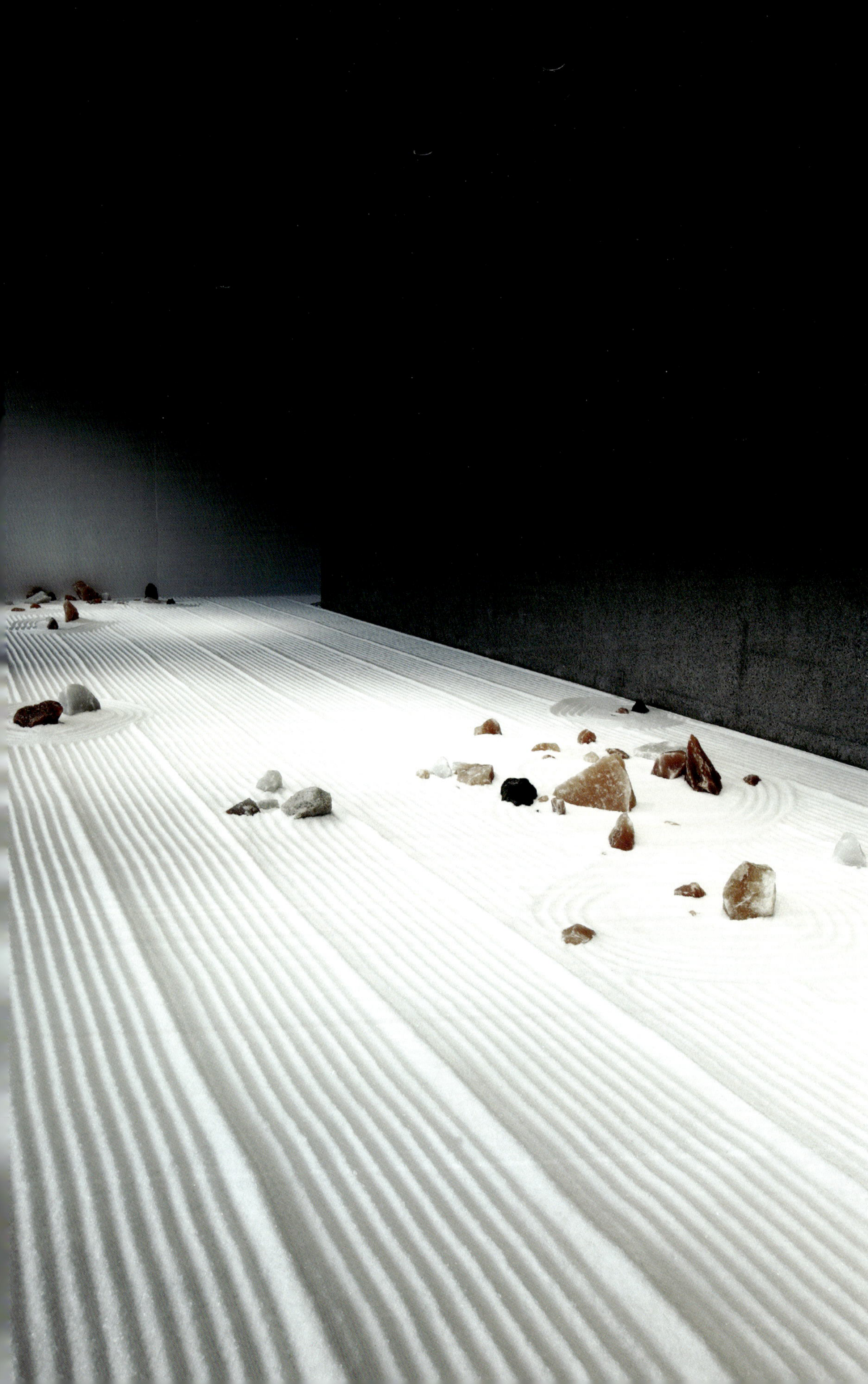

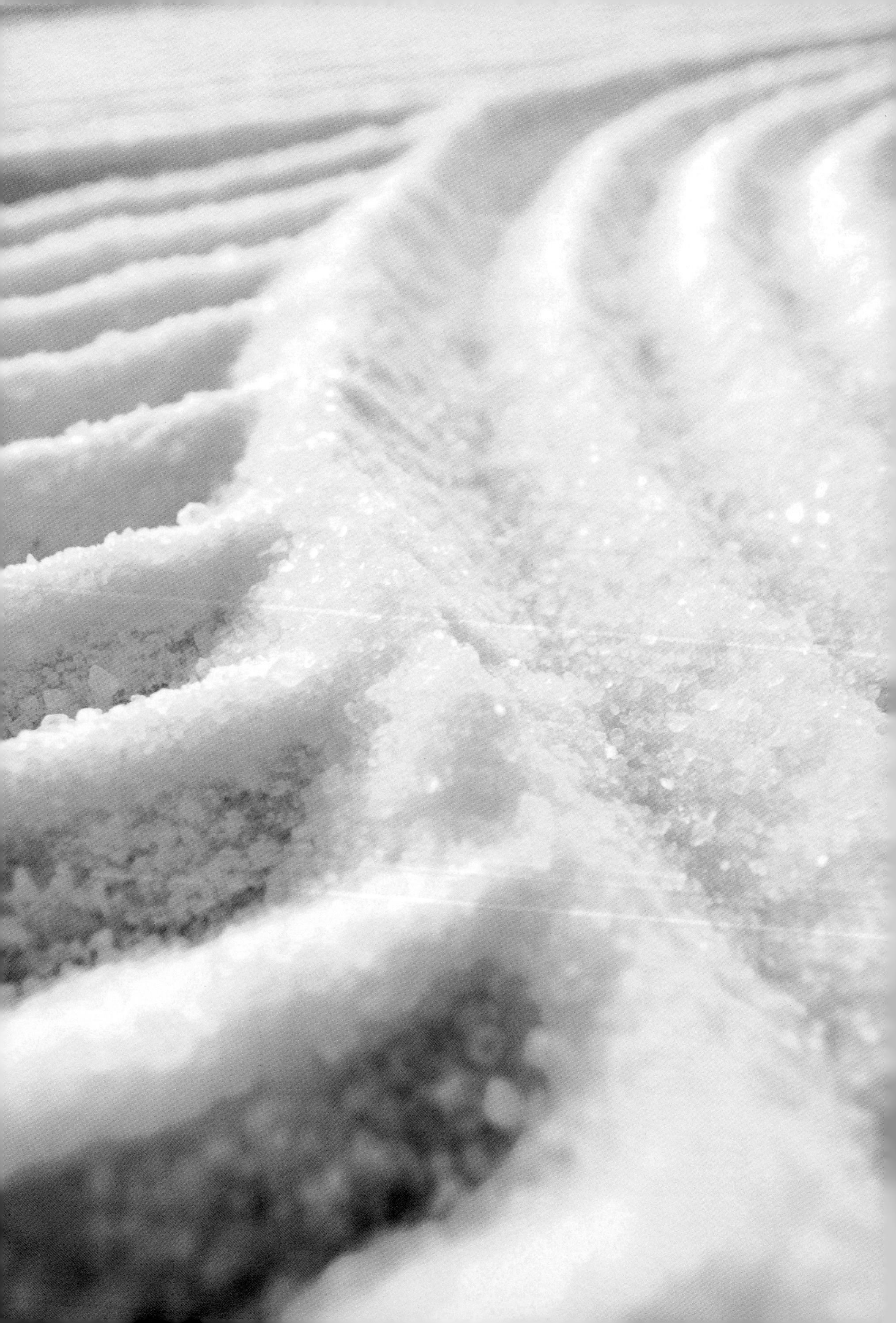

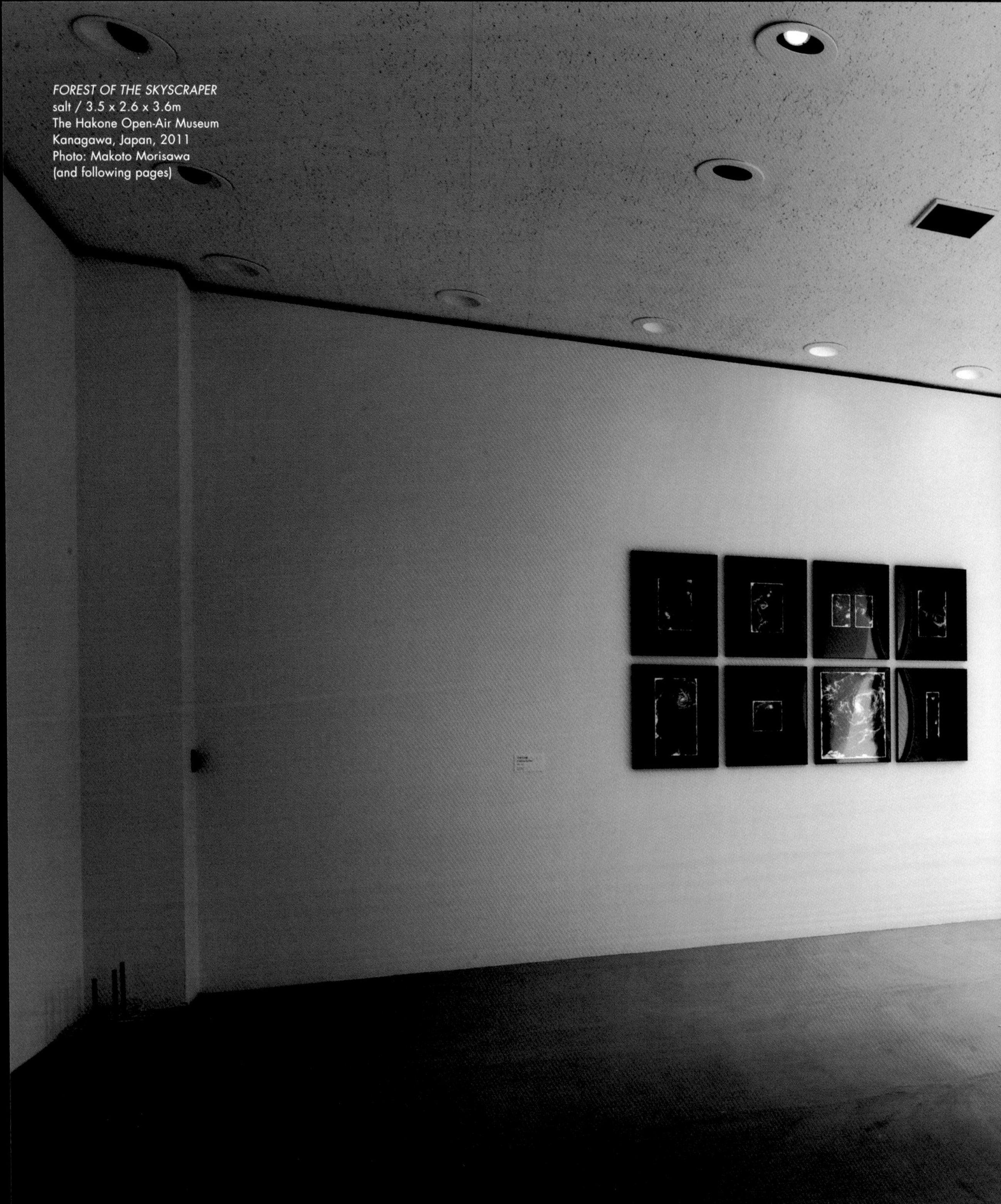

FOREST OF THE SKYSCRAPER
salt / 3.5 x 2.6 x 3.6m
The Hakone Open-Air Museum
Kanagawa, Japan, 2011
Photo: Makoto Morisawa
(and following pages)

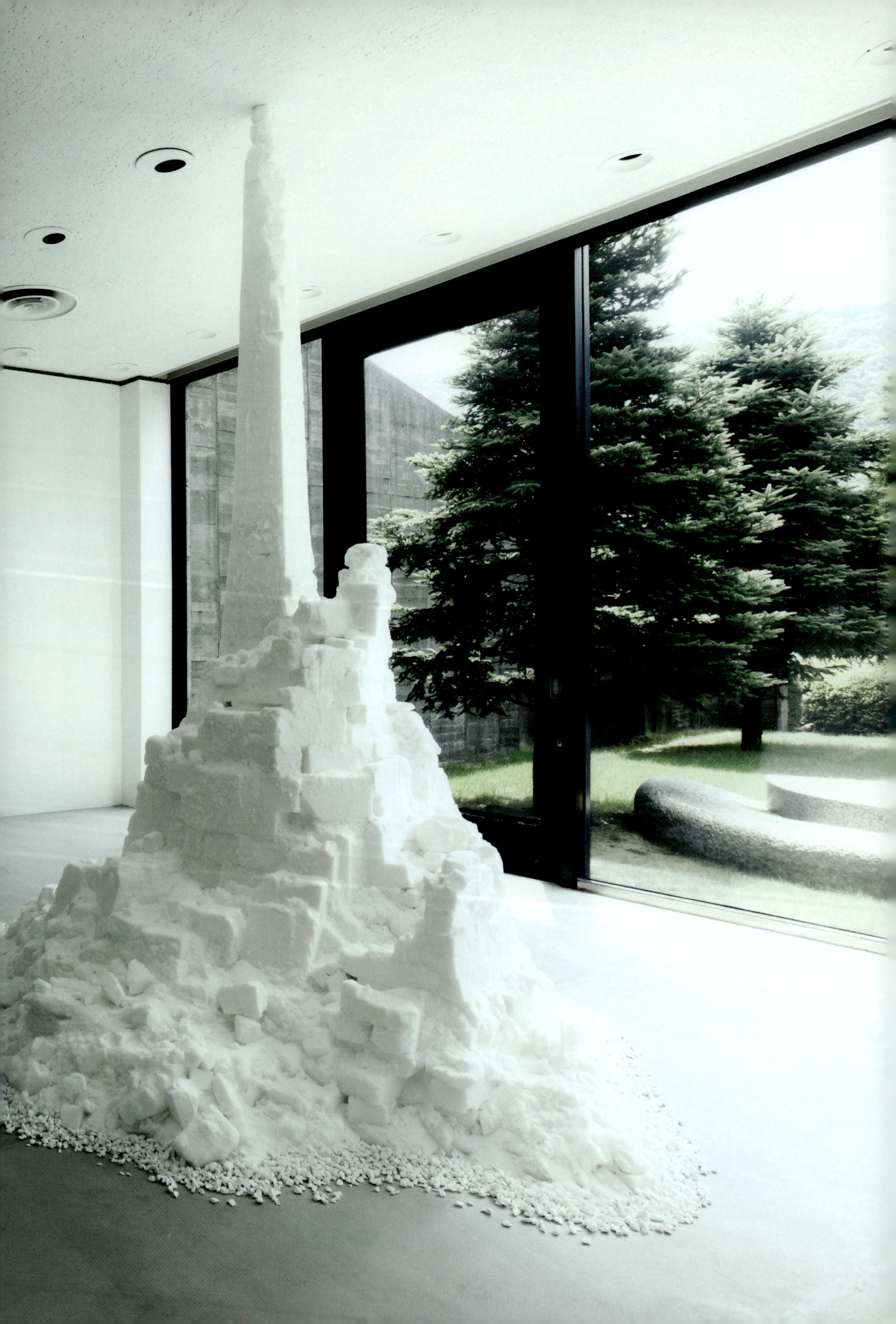

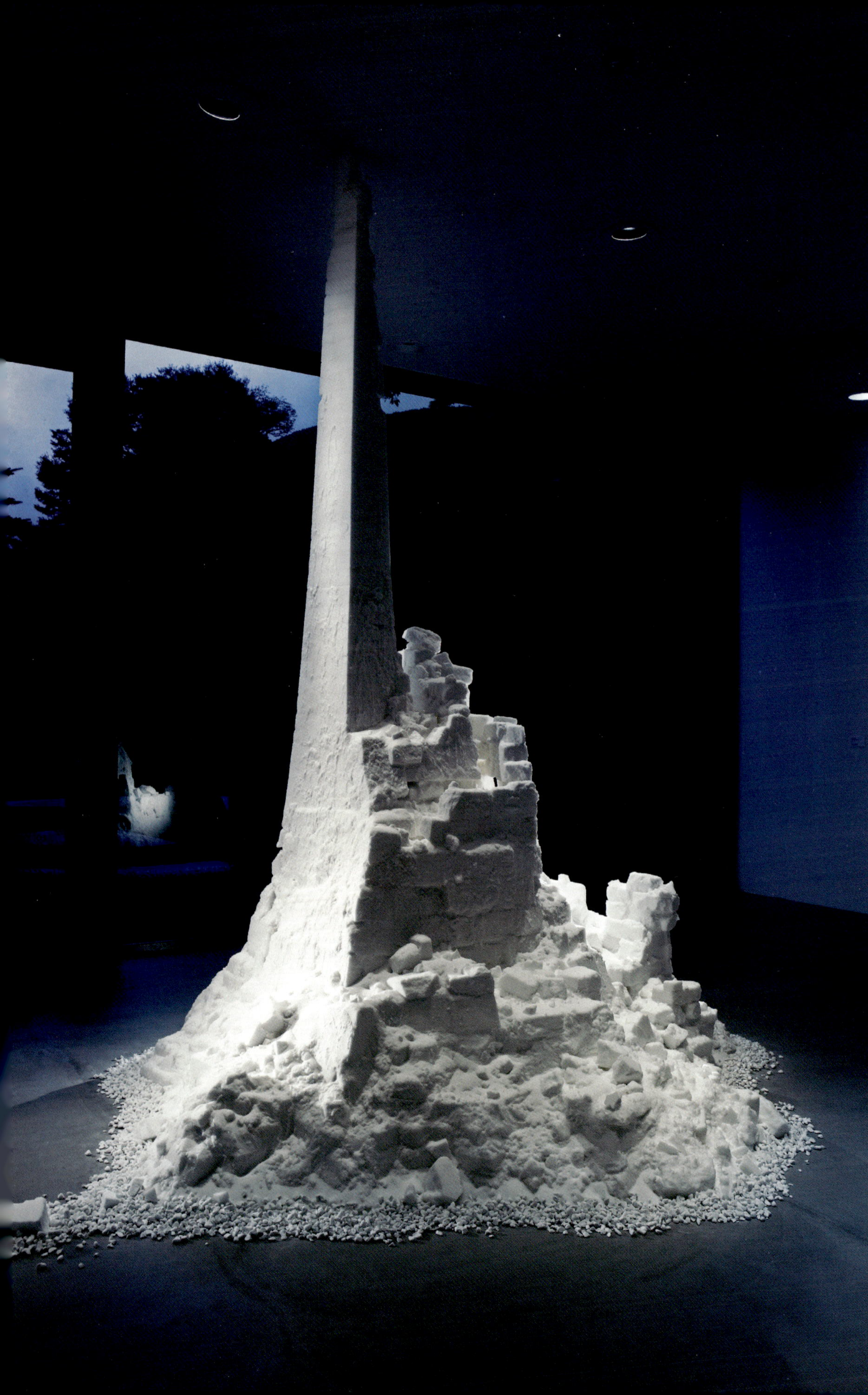

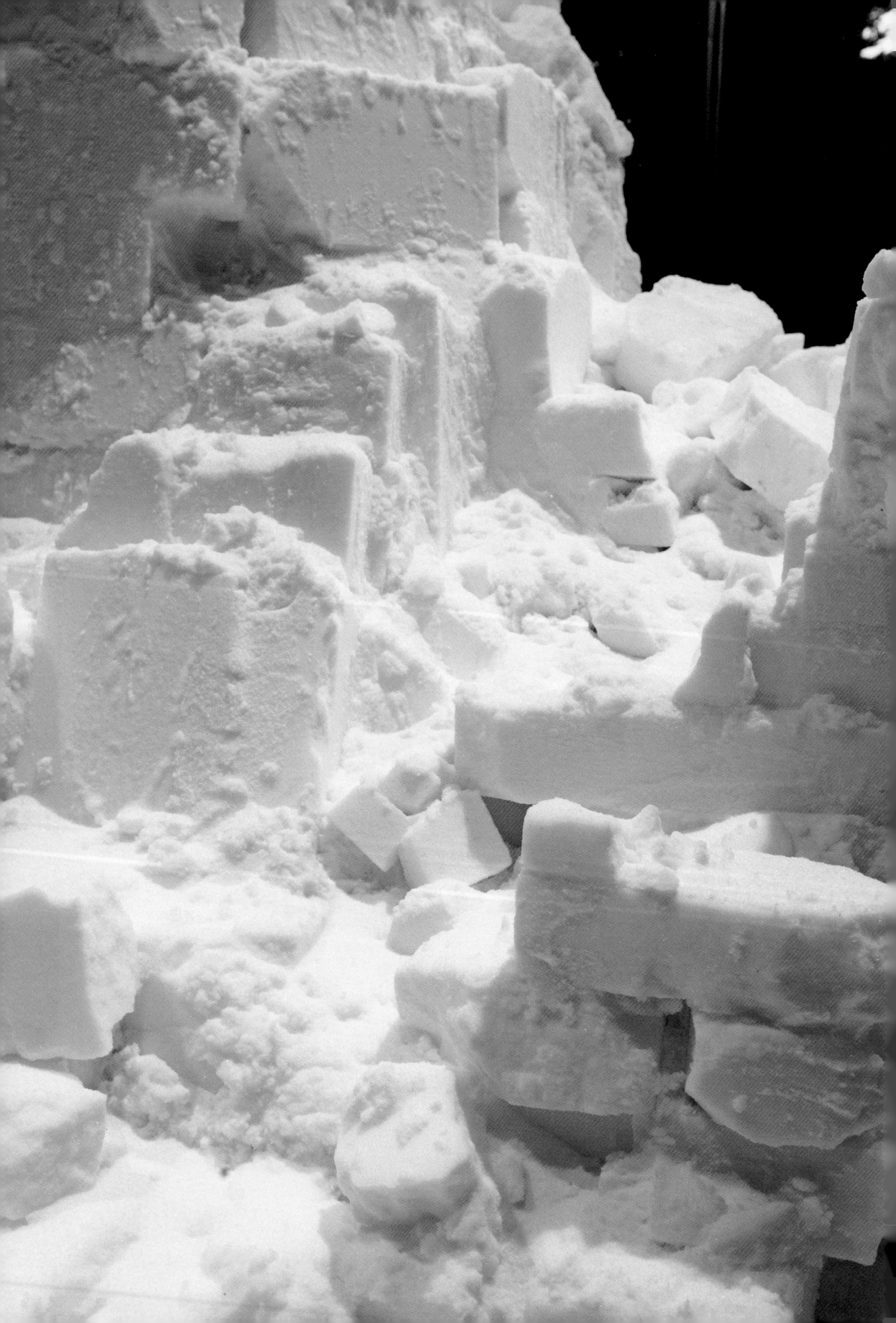

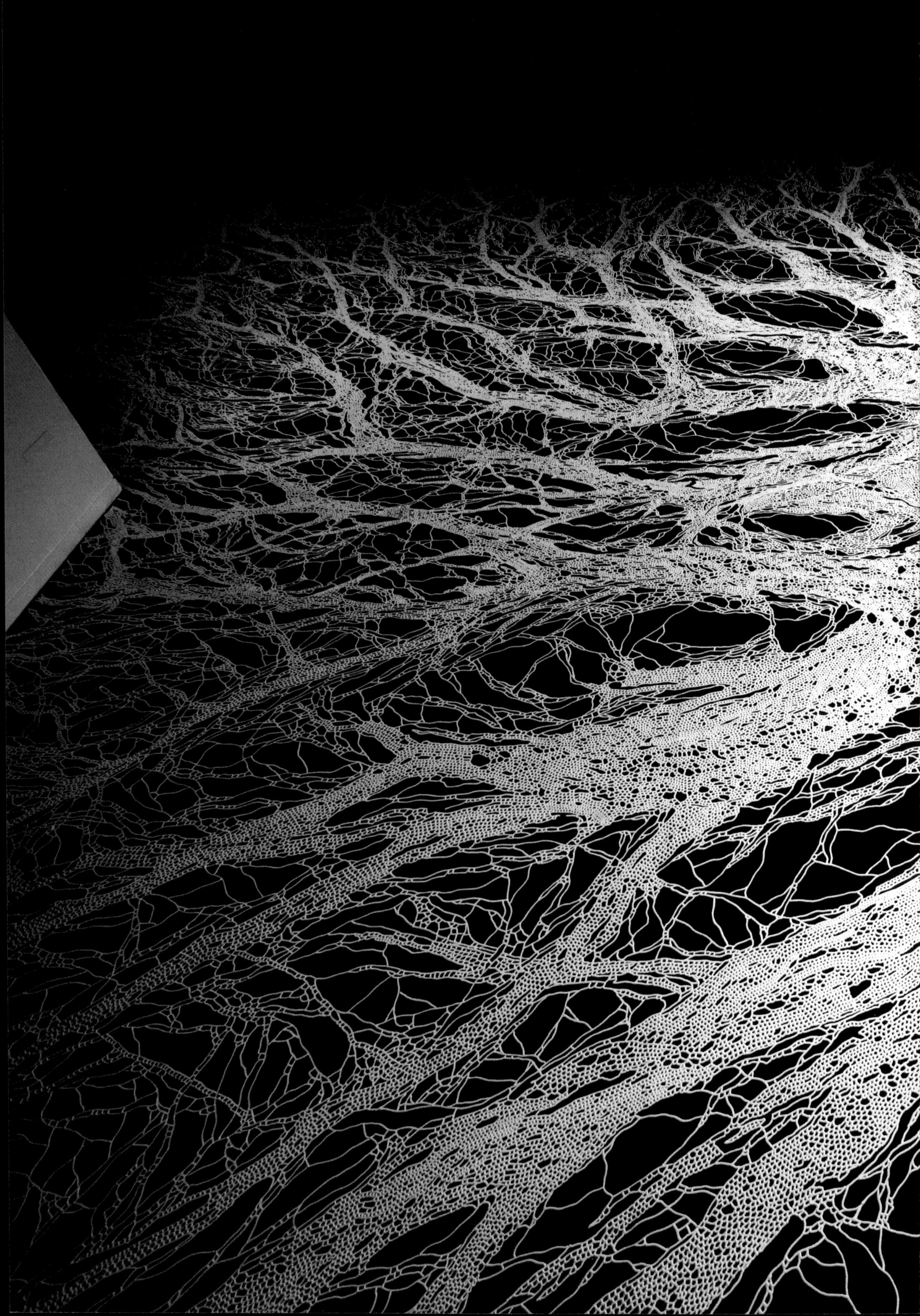

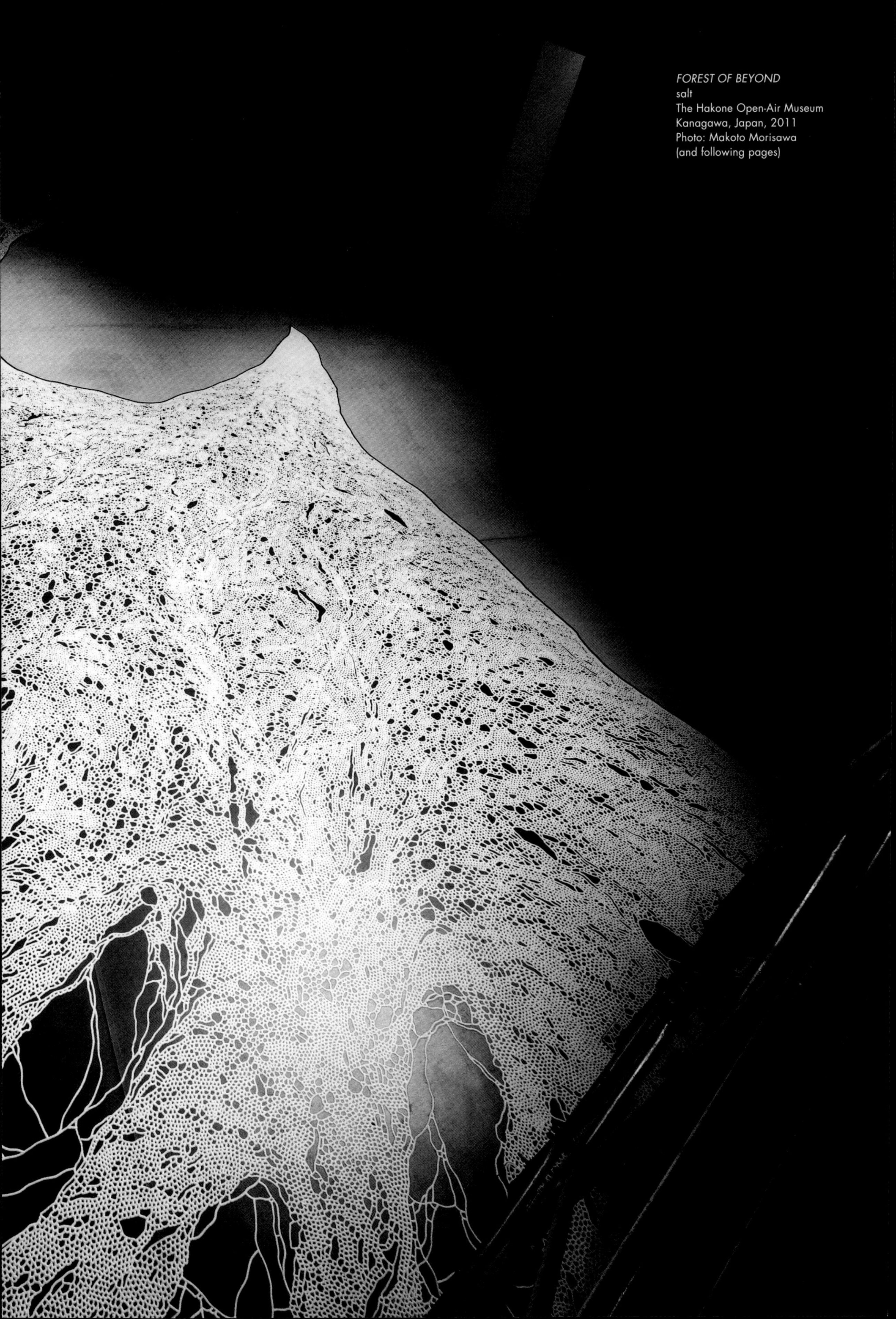

FOREST OF BEYOND
salt
The Hakone Open-Air Museum
Kanagawa, Japan, 2011
Photo: Makoto Morisawa
(and following pages)

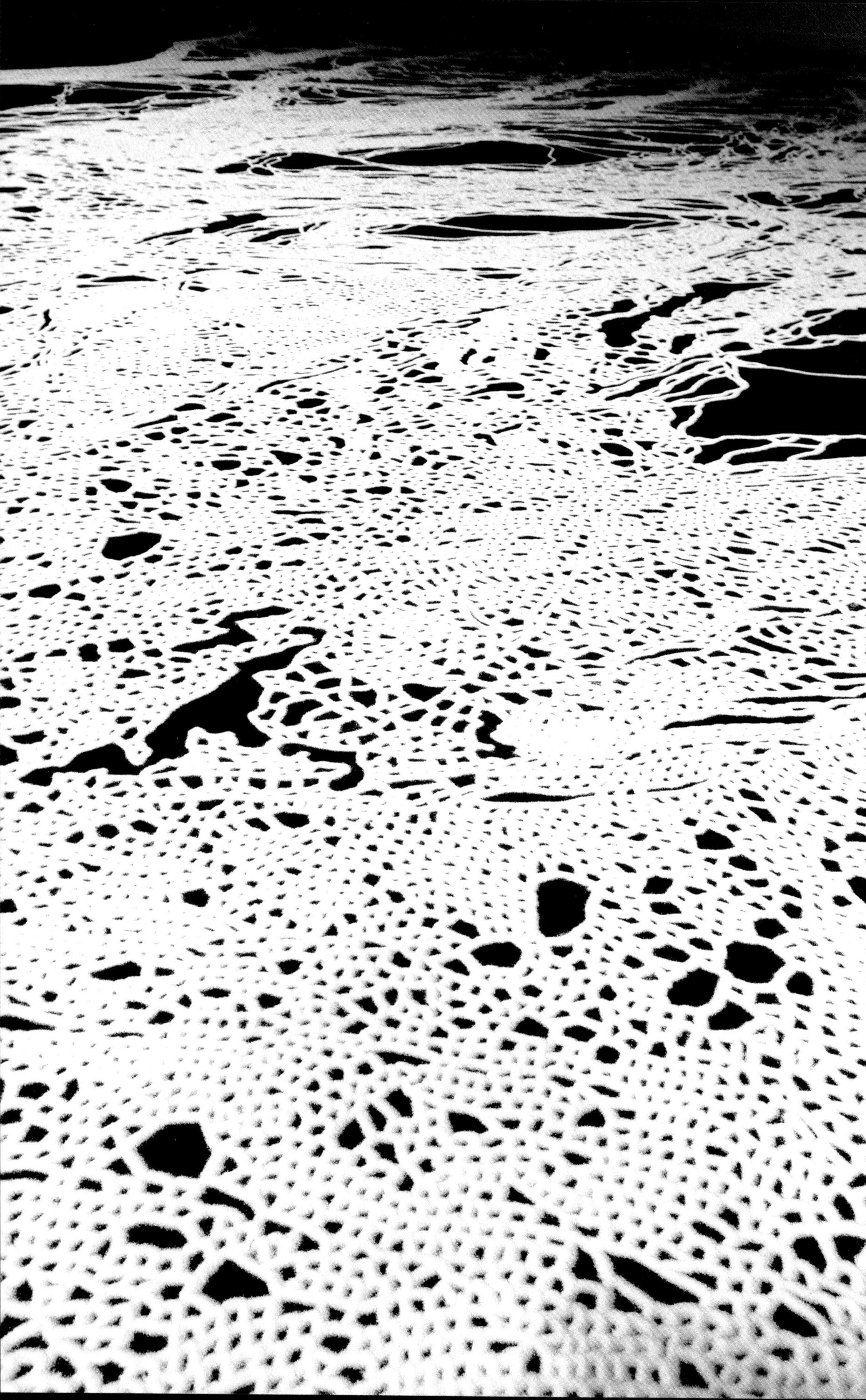

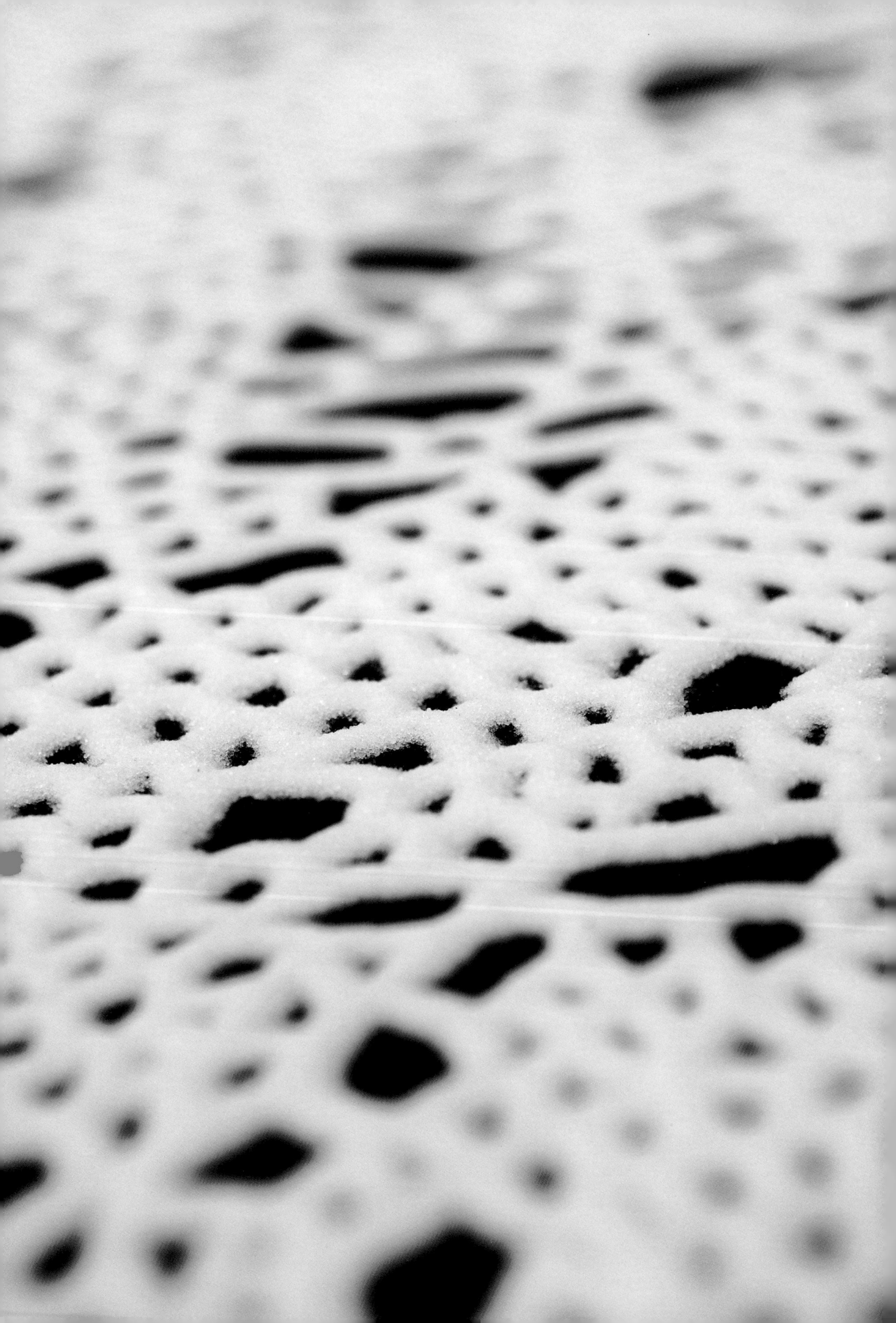

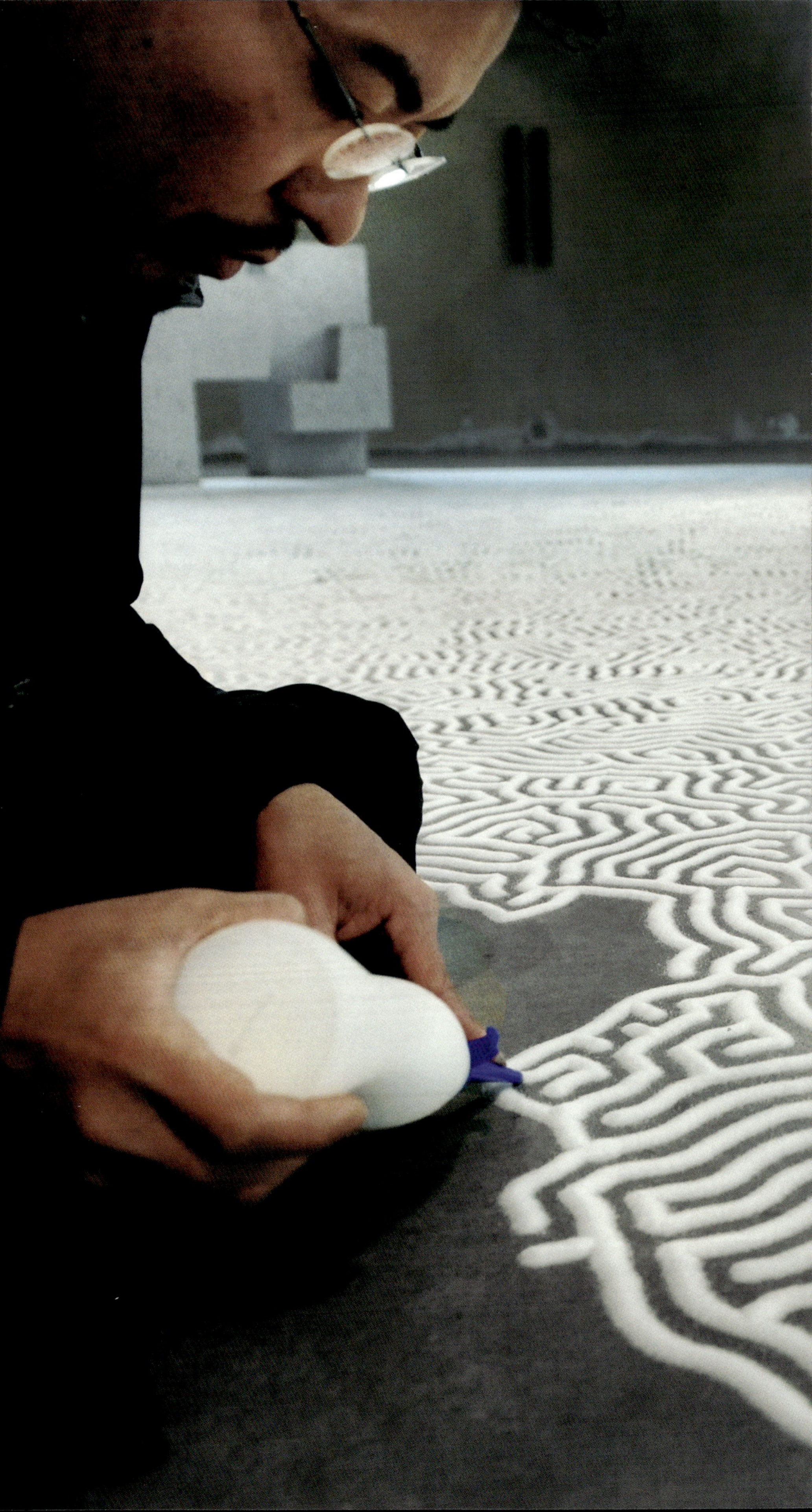

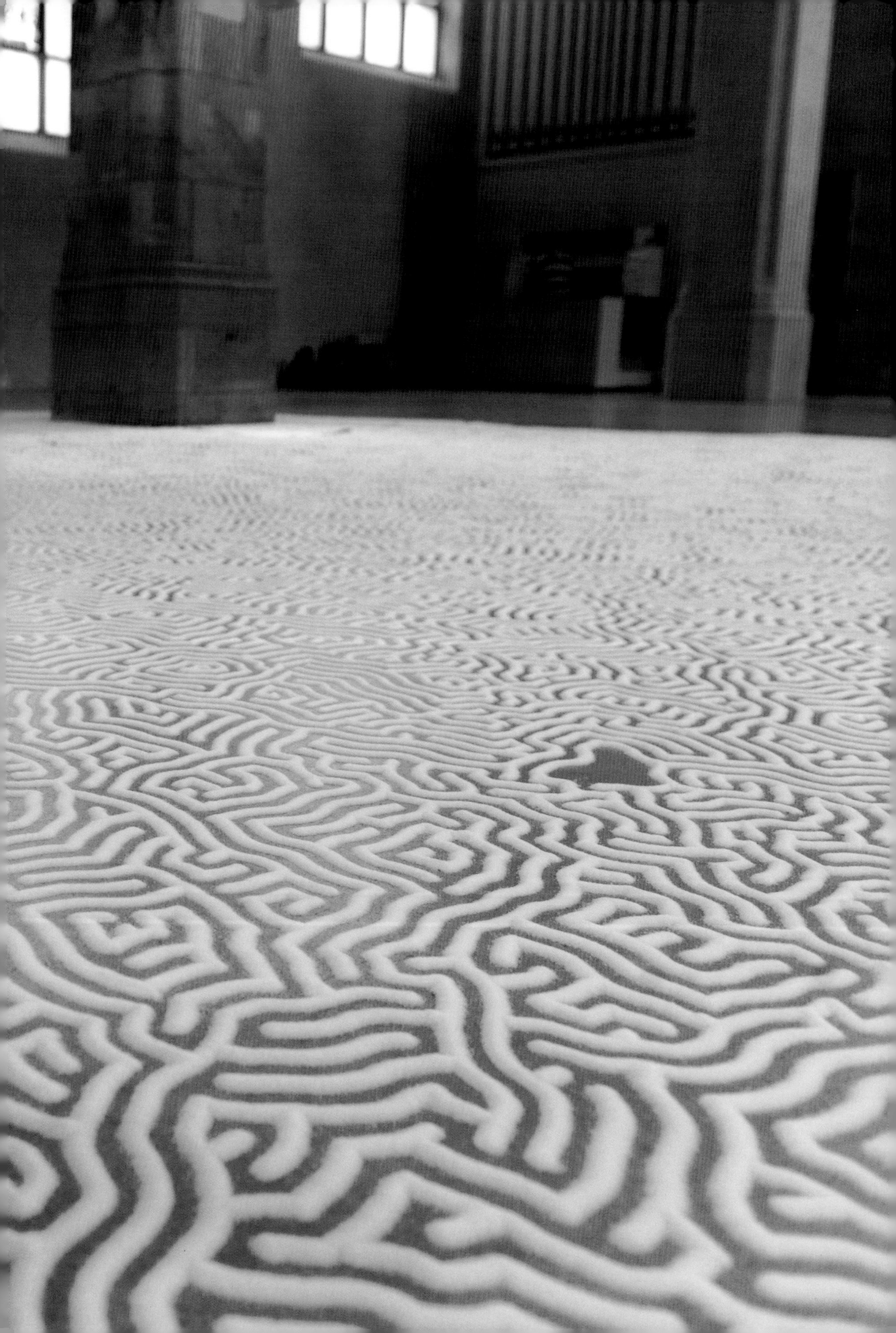

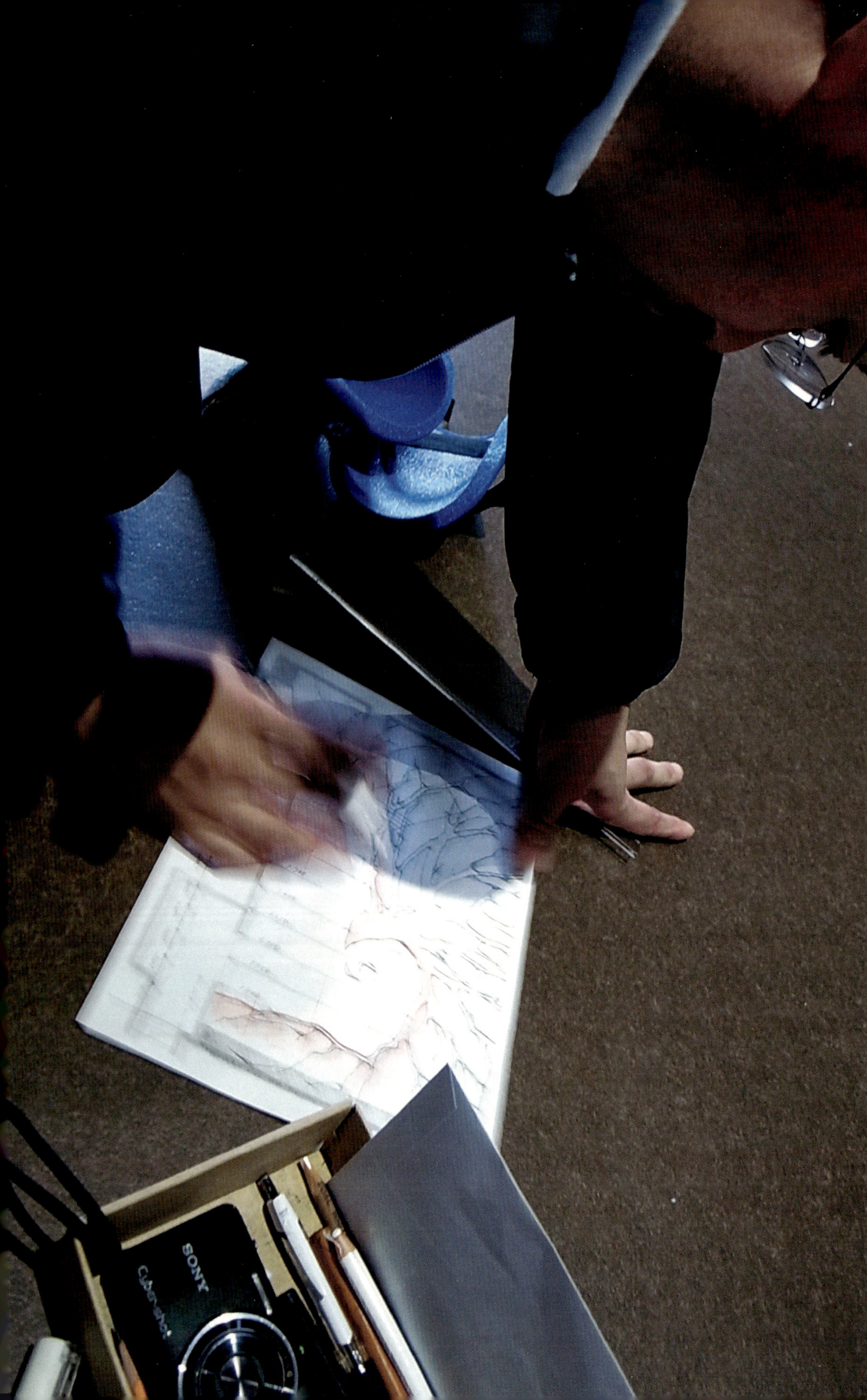
SONY
Cyber-shot

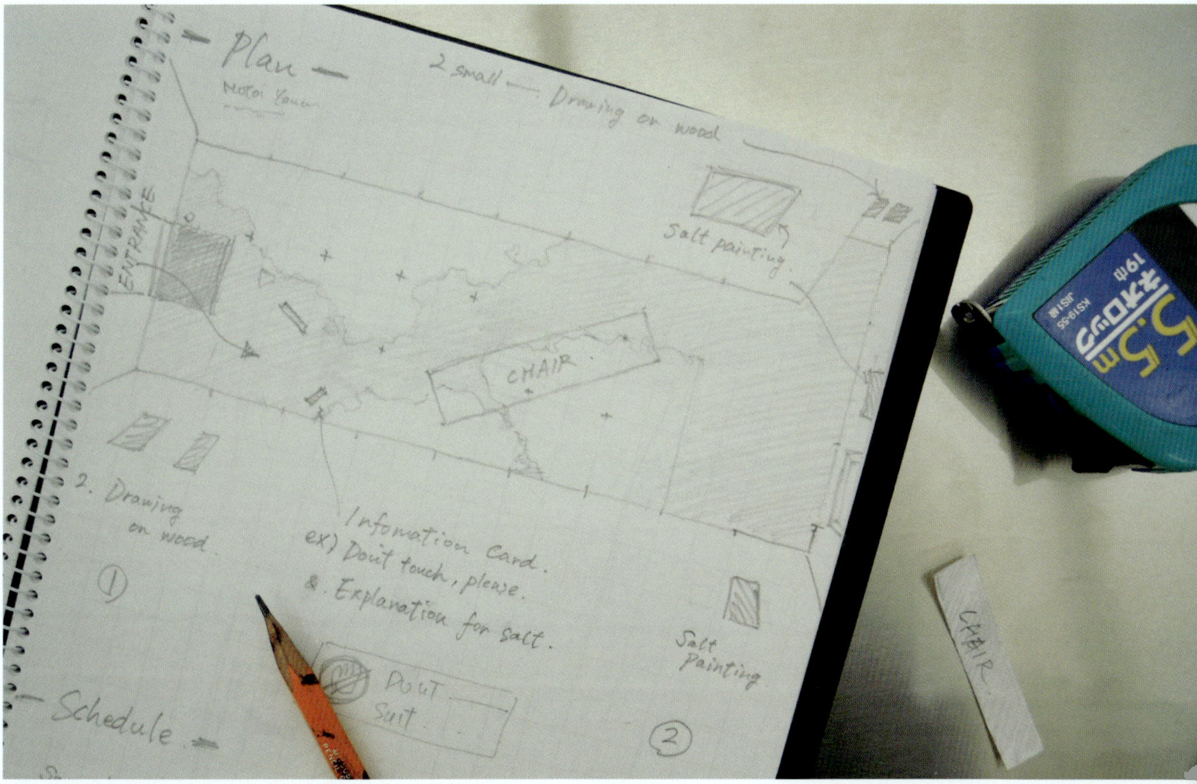
Plan
2 small — Drawing on wood
Salt painting
ENTRANCE
CHAIR
2. Drawing on wood
Infomation card.
ex) Don't touch, please.
&. Explanation for salt.
Salt Painting
Schedule
CHAIR

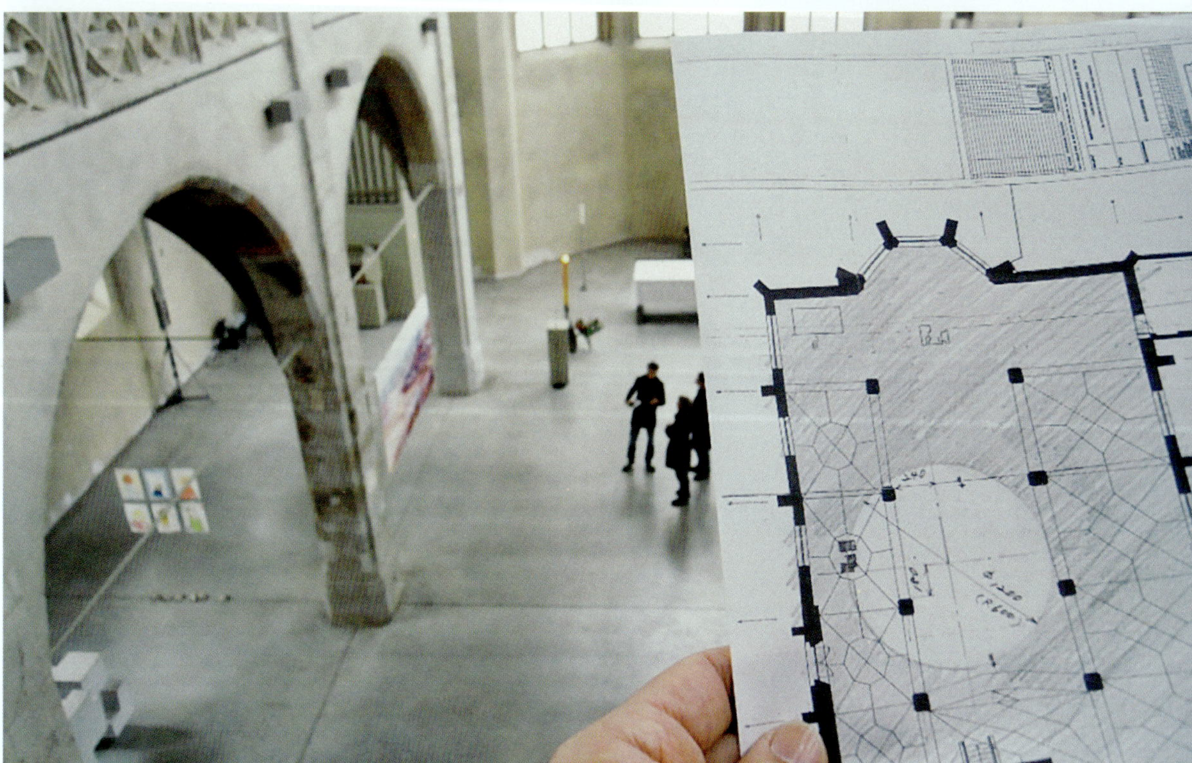

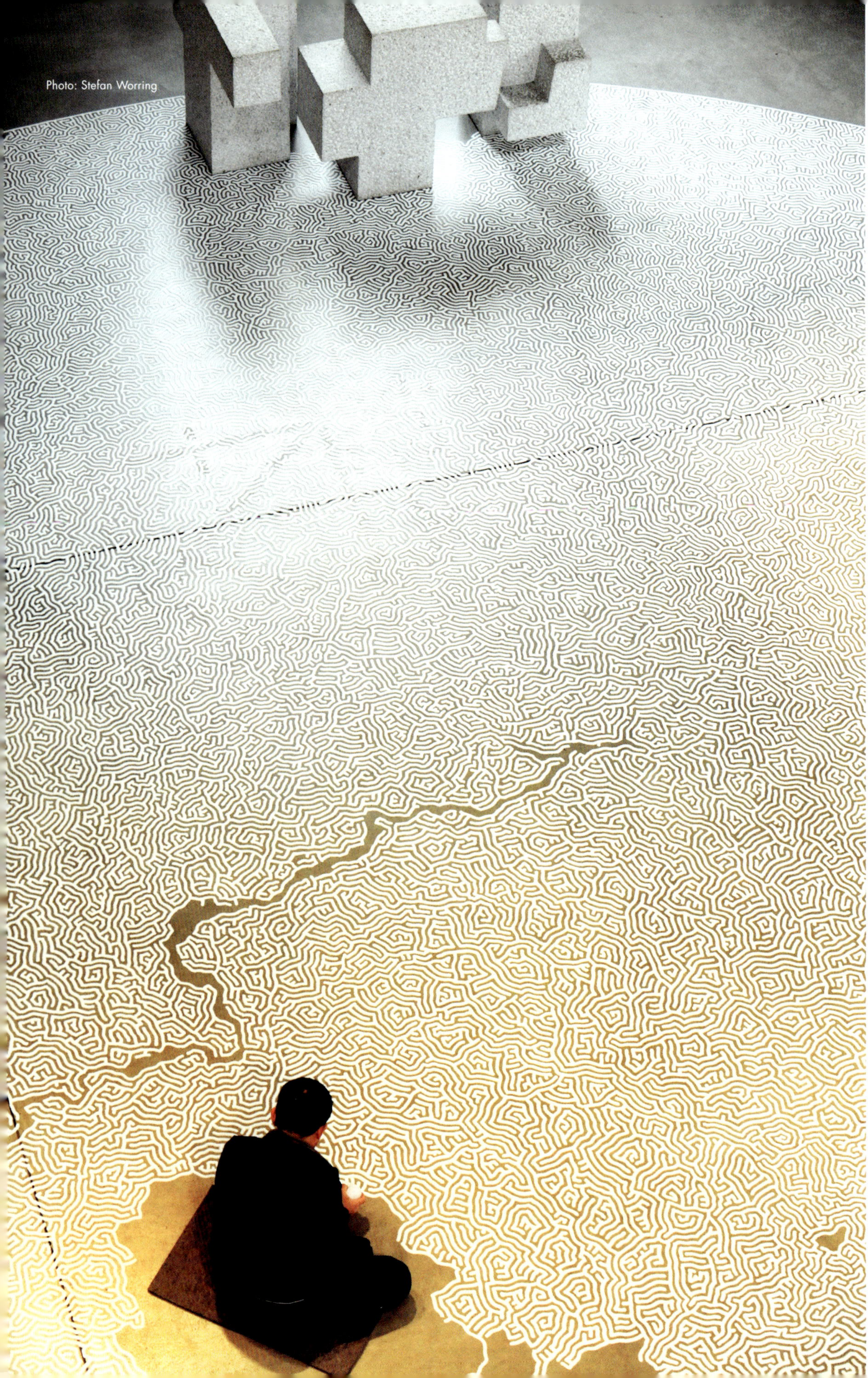

Photo: Stefan Worring

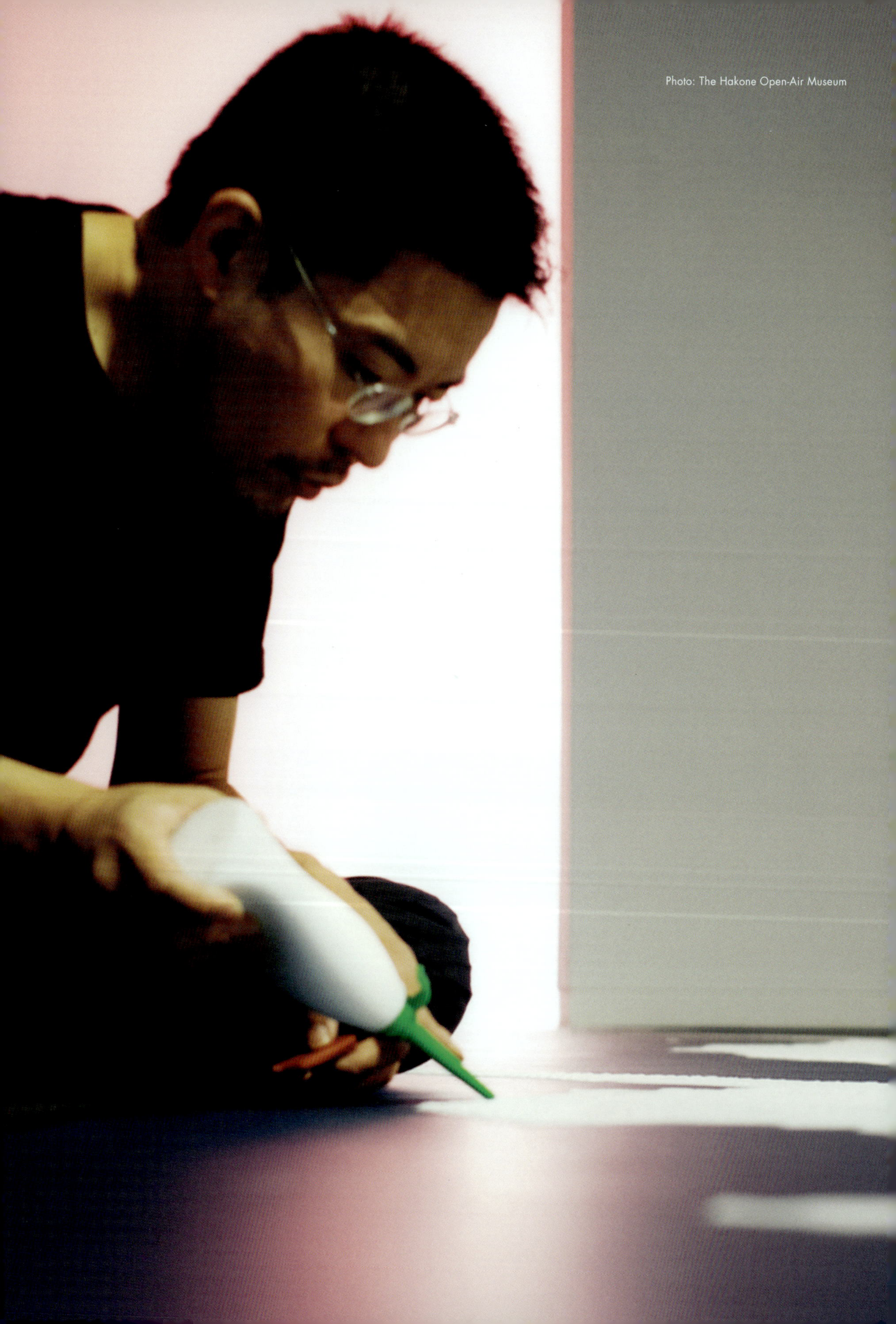

Photo: The Hakone Open-Air Museum

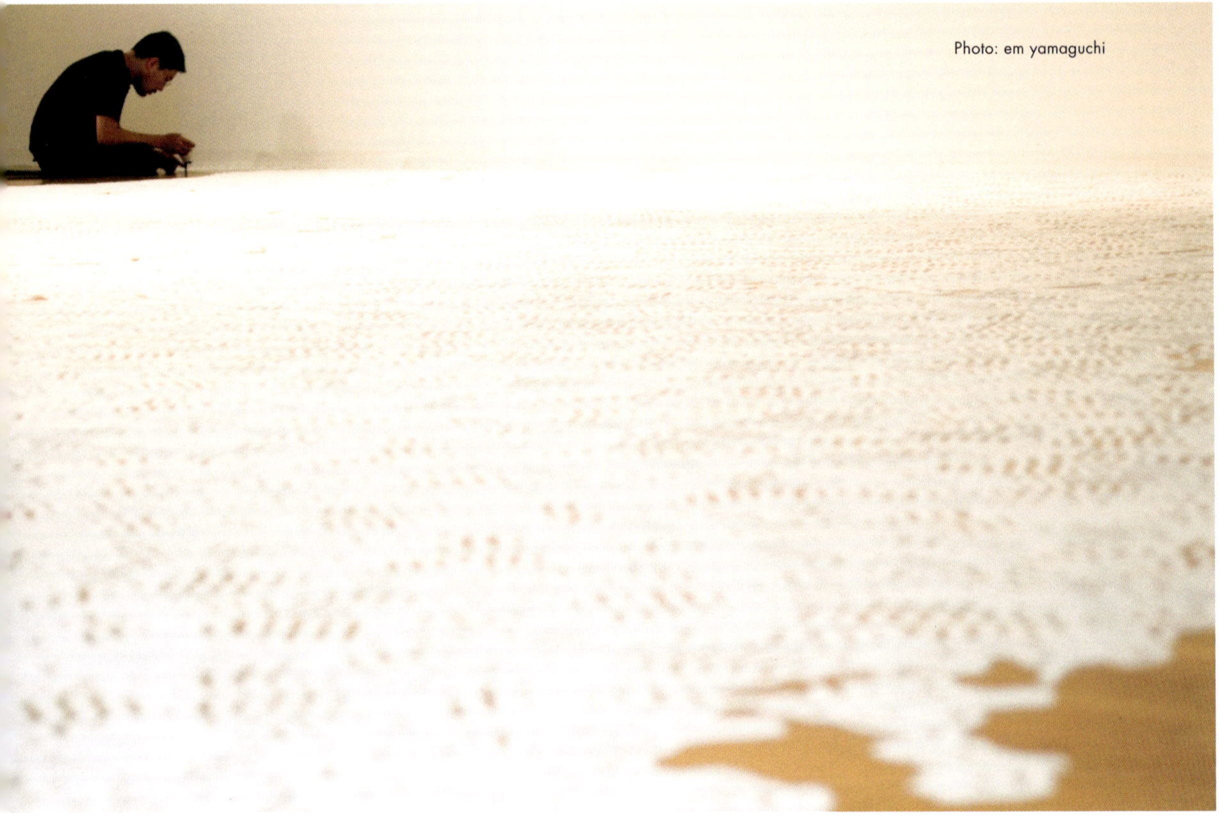
Photo: em yamaguchi

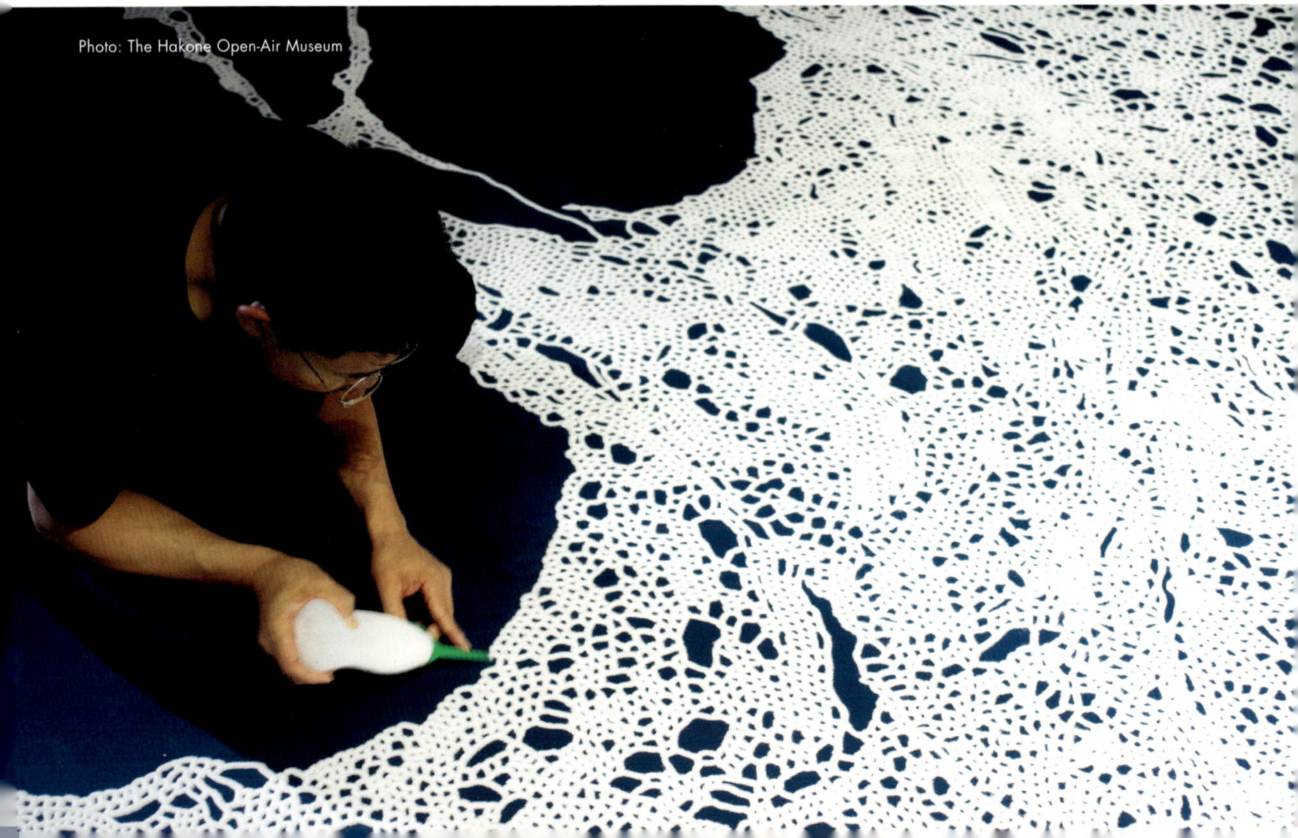
Photo: The Hakone Open-Air Museum

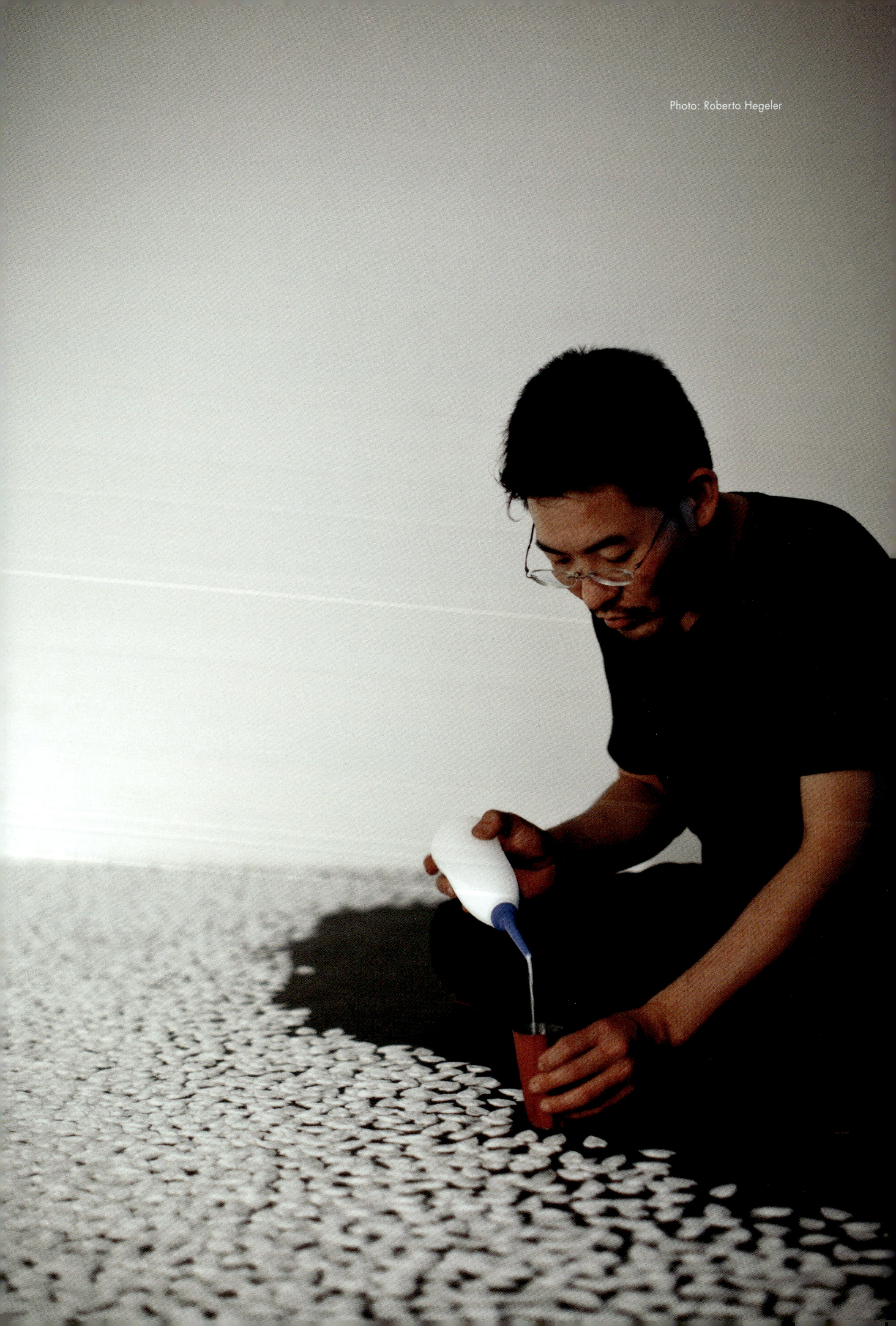

Photo: Roberto Hegeler

Return to the Sea

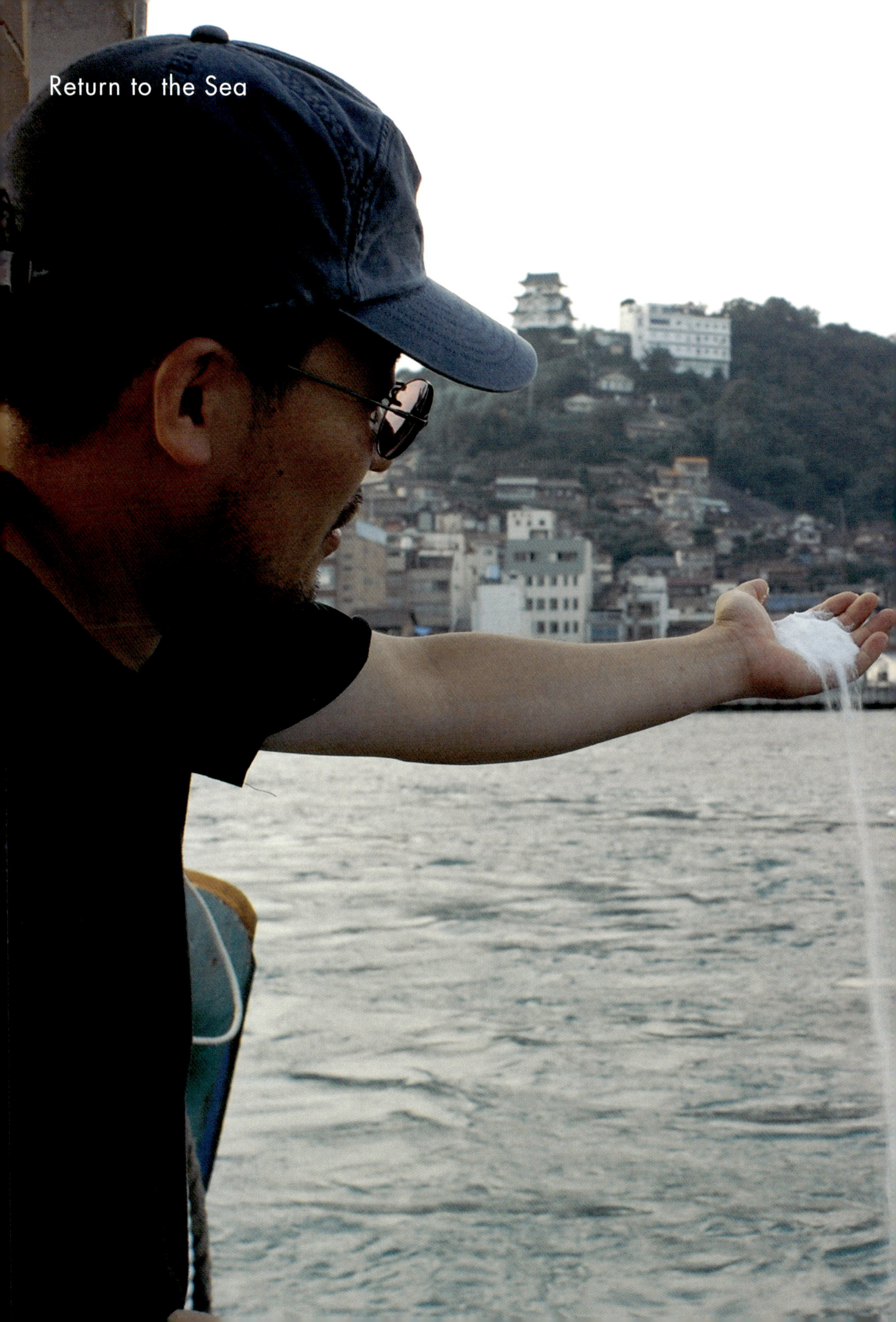

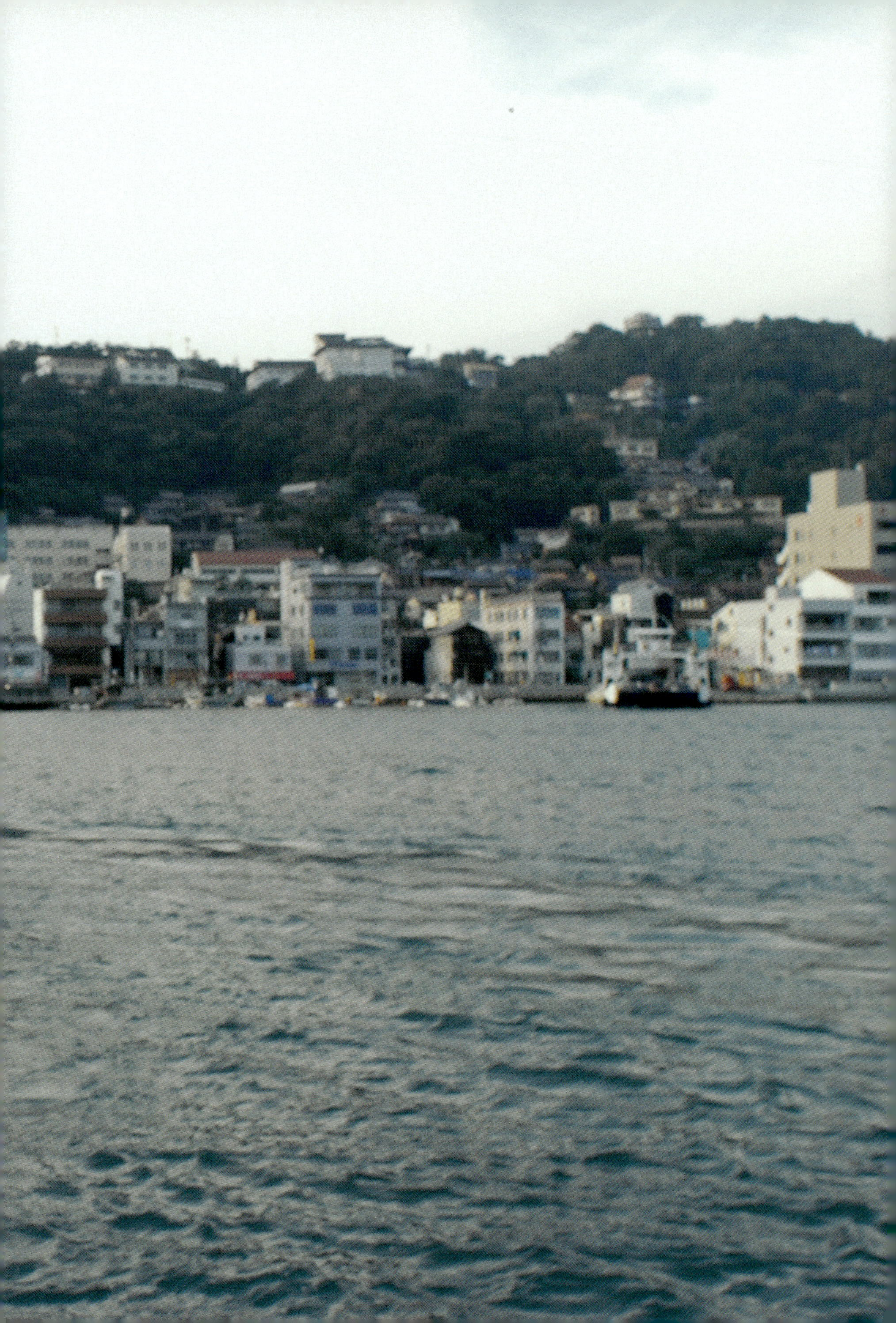

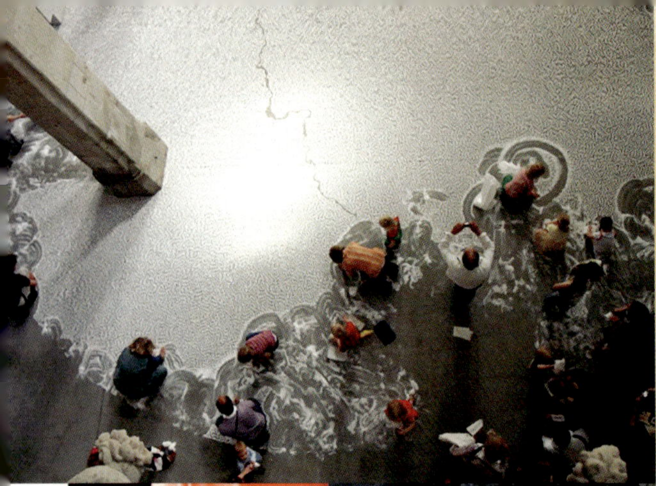
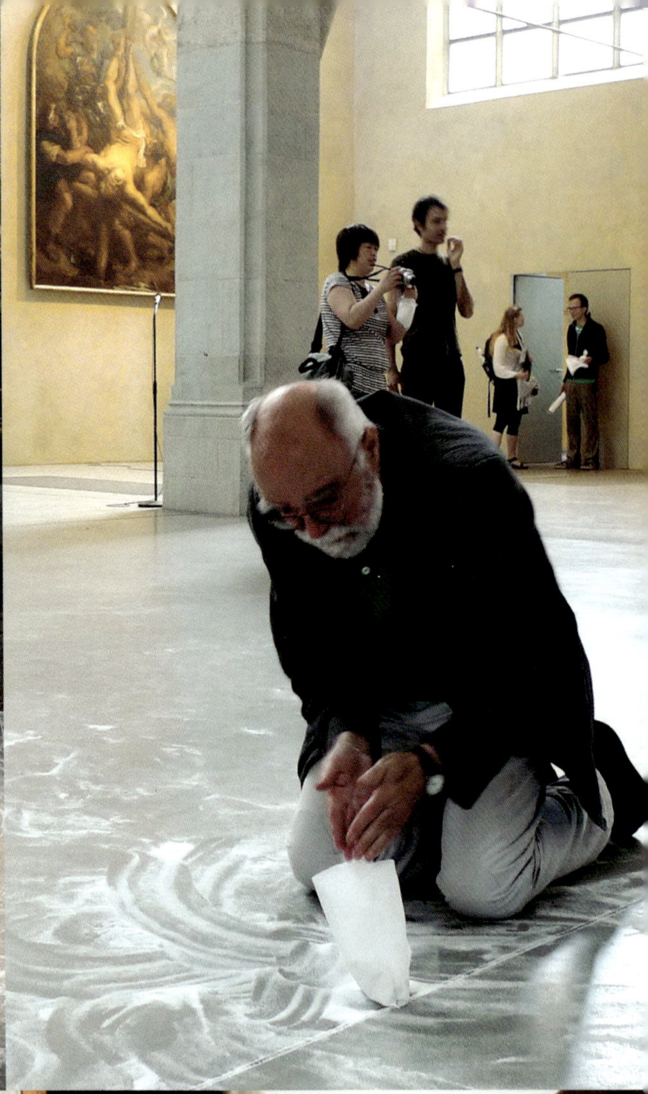
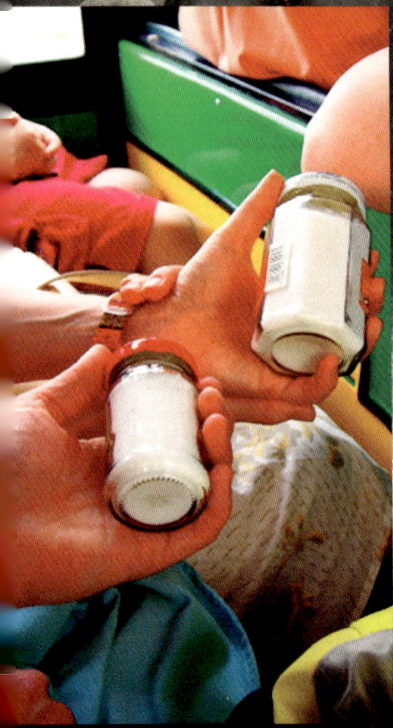
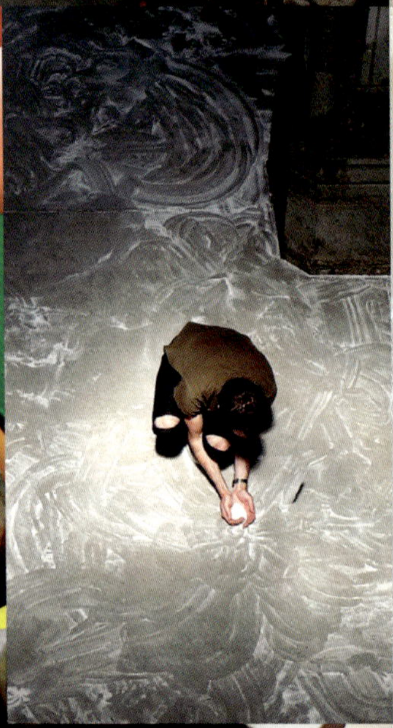
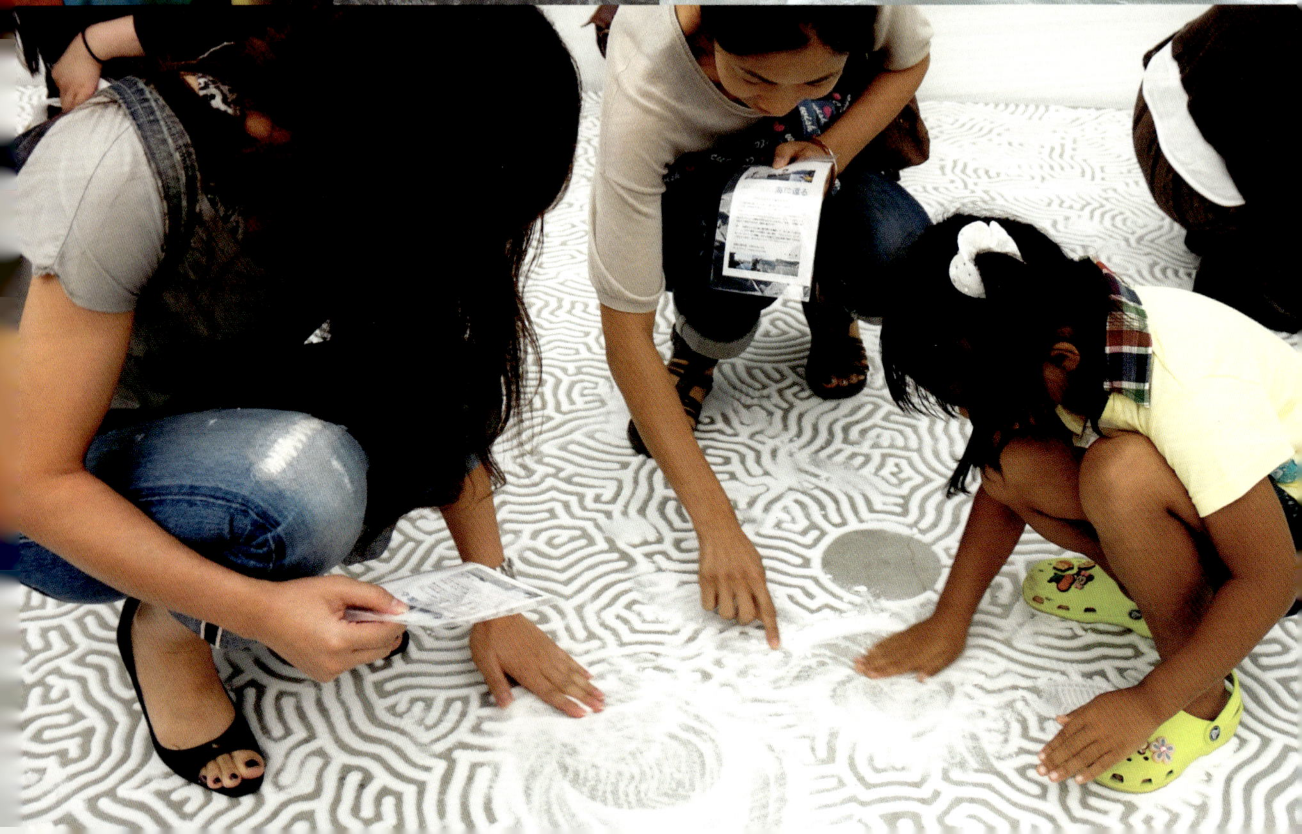

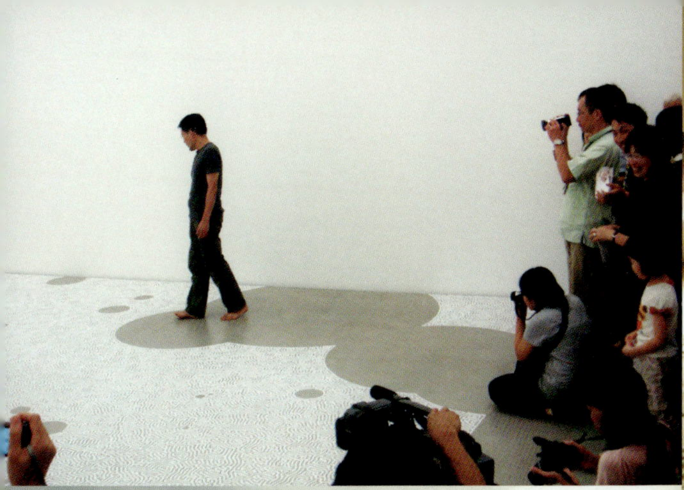
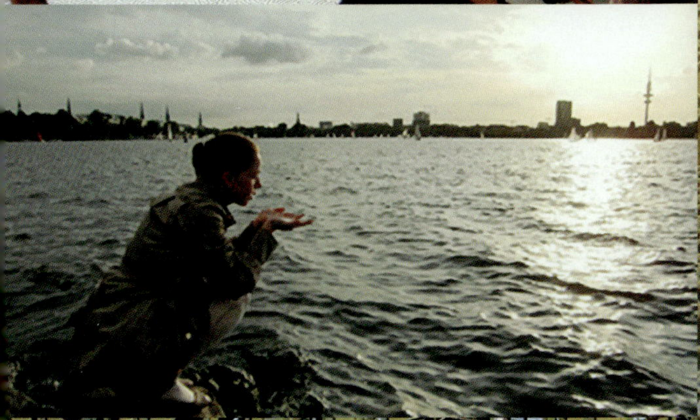
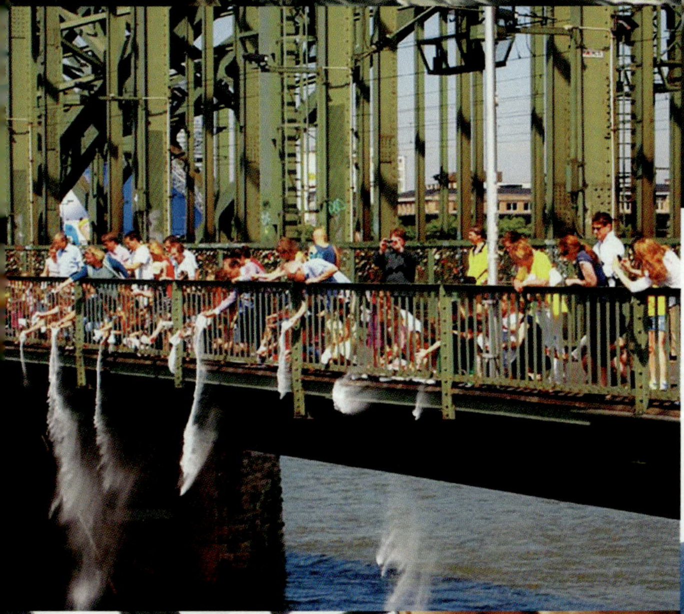
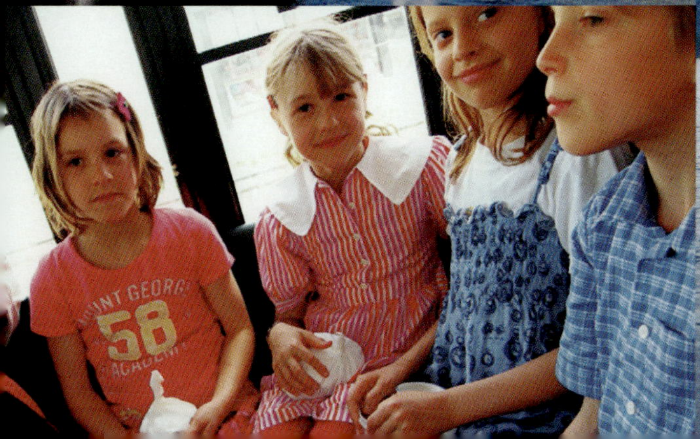
58
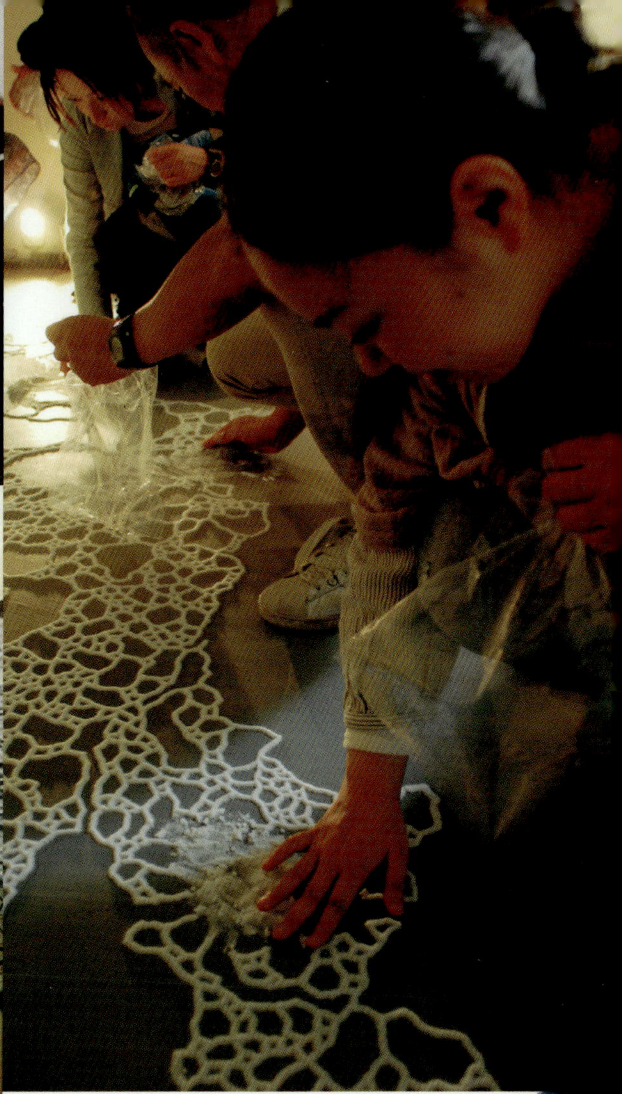
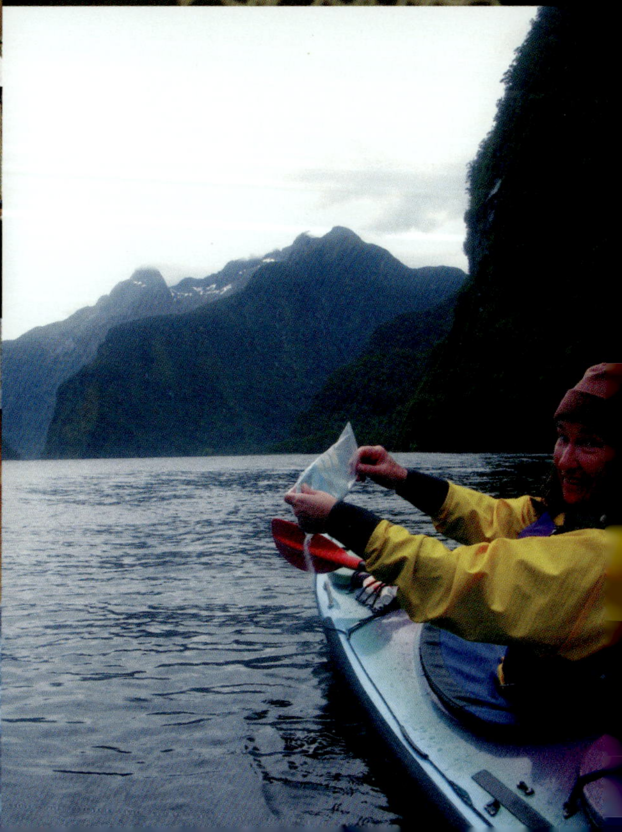

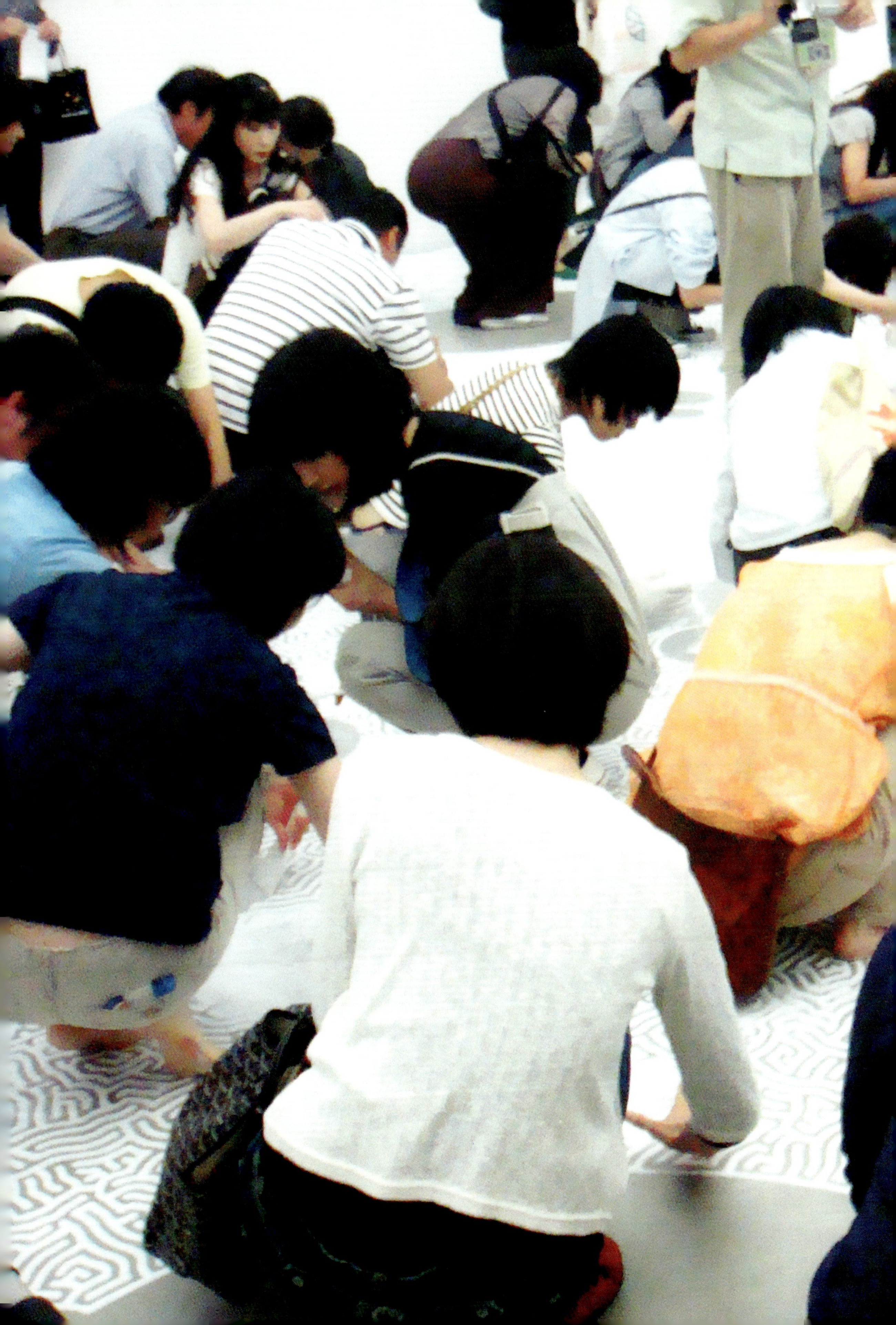

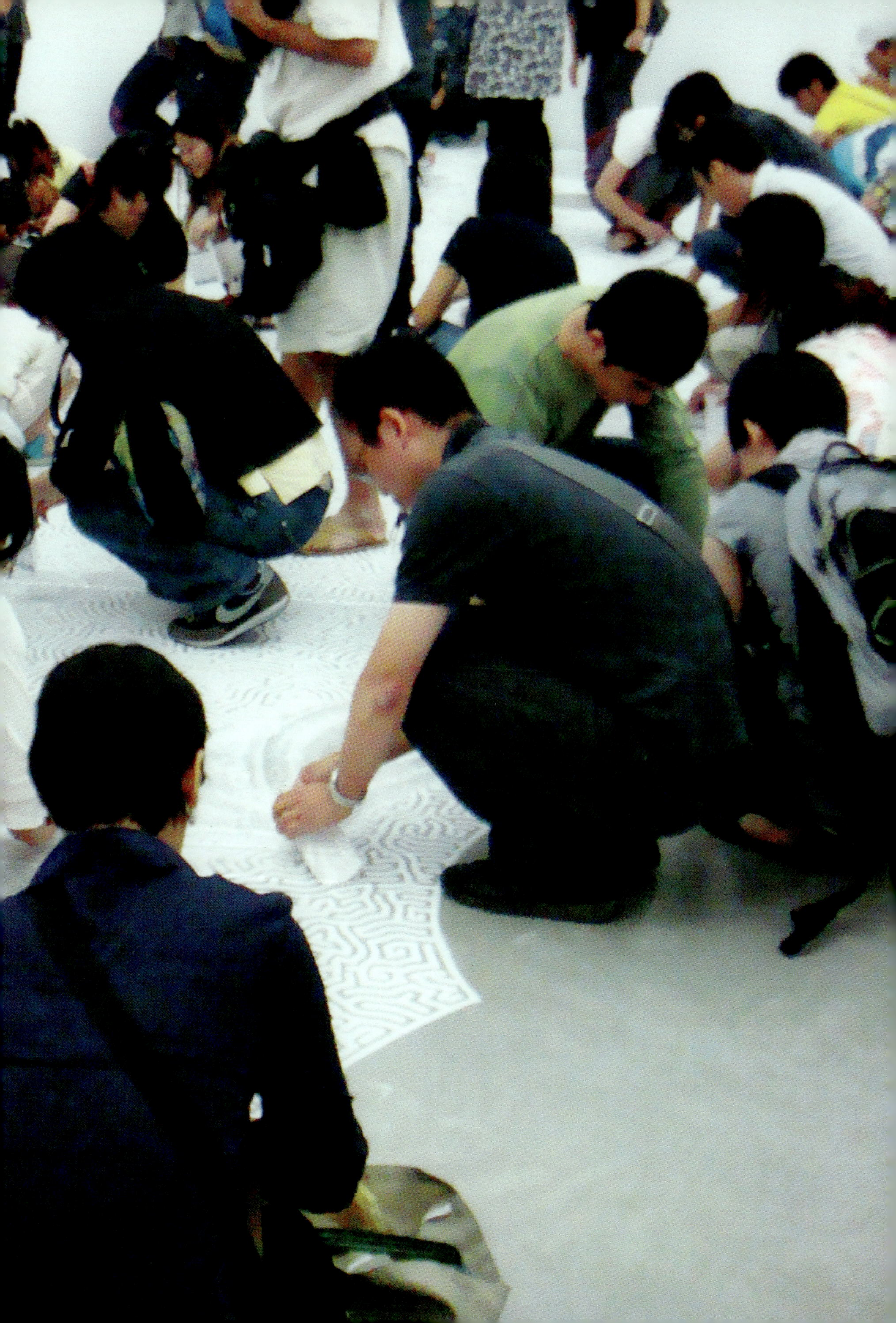

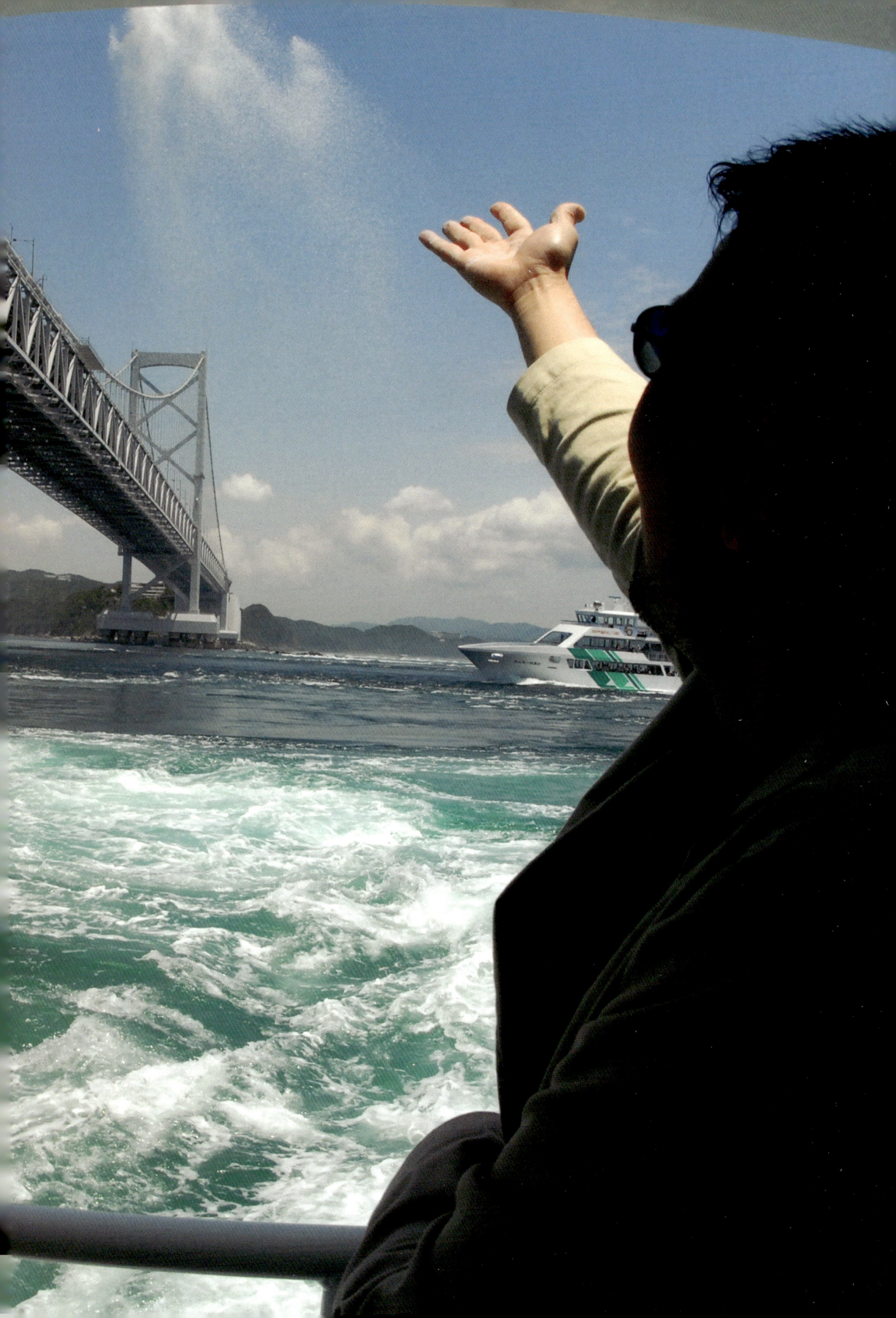

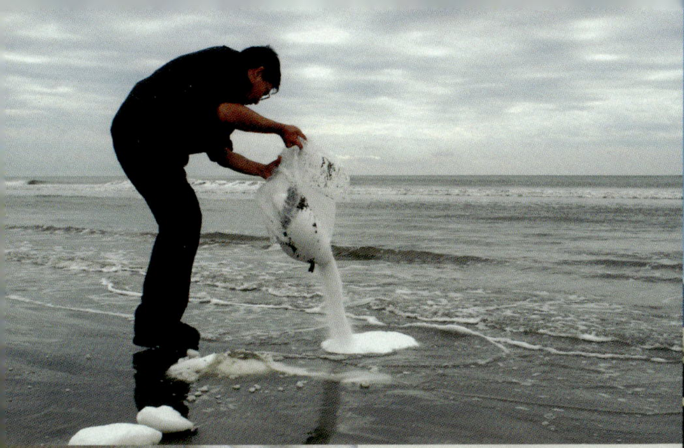
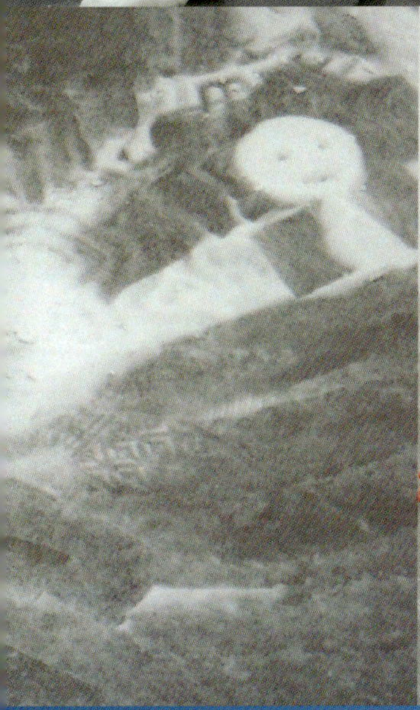
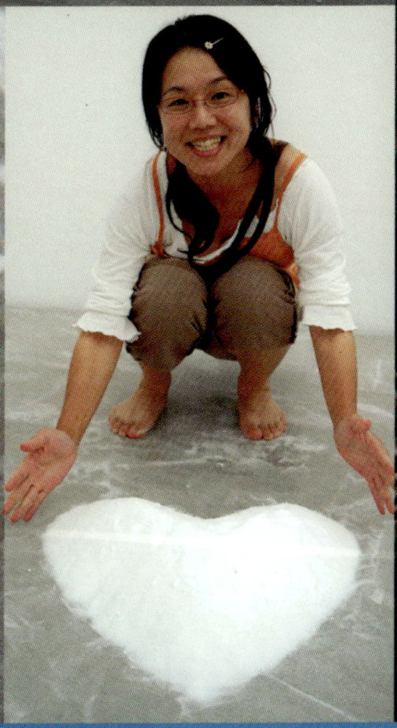
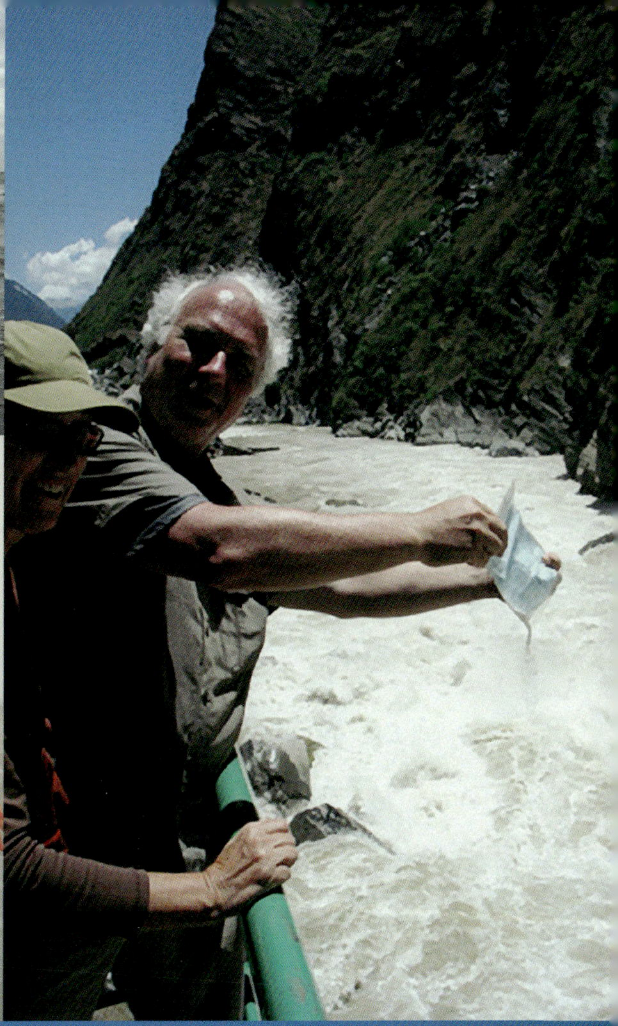
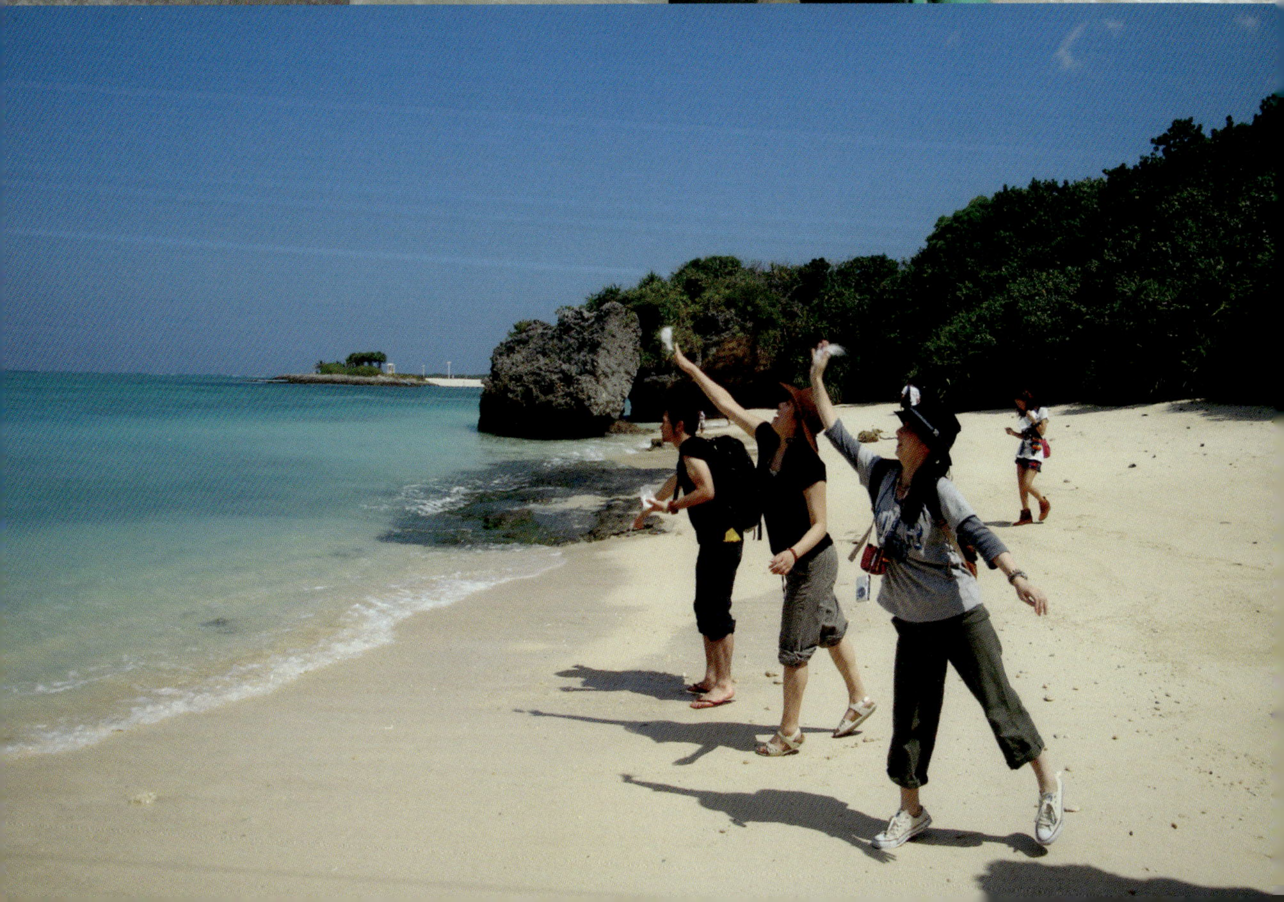

The True Value of Salt

Mark Kurlansky

"A major event in the history of the West," wrote Argentine poet Jorge Luis Borges, "was the discovery of the East." He pointed out that this was an ongoing process. Westerners discovered two things in the East: Asians live the same human saga as westerners, but they experience it differently.

Nothing better illustrates this East-West paradox between the same and different than the Japanese relationship to salt.

Salt is a universal part of human life, yet in Japanese culture it has its own peculiar, elegant traditions. Every nation and culture in the world has a salt story, either about producing and trading the mineral or about needing and acquiring it. China's story is that of a great producer, Japan's is of a small-scale producer for whom salt was always somewhat scarce and thus highly valued. Most of Japanese history is a story of scarcity and the great value placed upon simple things. And so in Japan, salt, like most precious things, became an important cultural symbol.

Salt lends itself to metaphor. It has many mysterious properties: it preserves food, makes objects float, and puts out fires. Depending on how salt is used, it both lowers temperature, as in refrigeration, freezing, and ice cream making, and raises it, as in deicing roads, which is the contemporary world's leading use of salt.

Even in chemistry, salt lends itself to metaphor. The chemical definition of a salt is "an acid that acts on a base." There are numerous acids that act on numerous bases. It is the fusion of two very volatile elements, one unstable, because it is missing an electron, and one precarious, because of an extra electron. The two complement and complete each other as in a good marriage, and result in an extremely stable compound—a salt.

The salt that we eat is a compound of sodium and chloride. Sodium, an explosive metal, is never found by itself in nature. Chloride, alone, is a poisonous gas.

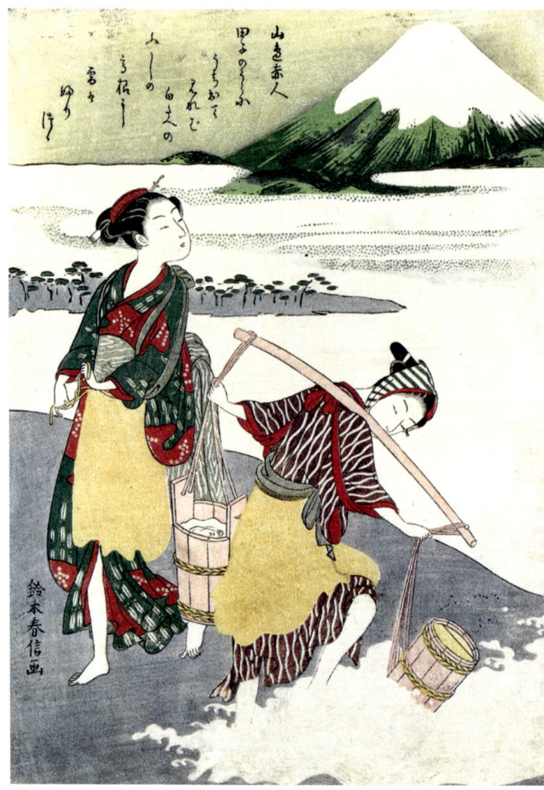

Salt Maidens on the Tago-no-ura Beach with Mt. Fuji Behind,
by Suzuki Harunobu (1725–70)
color woodblock print

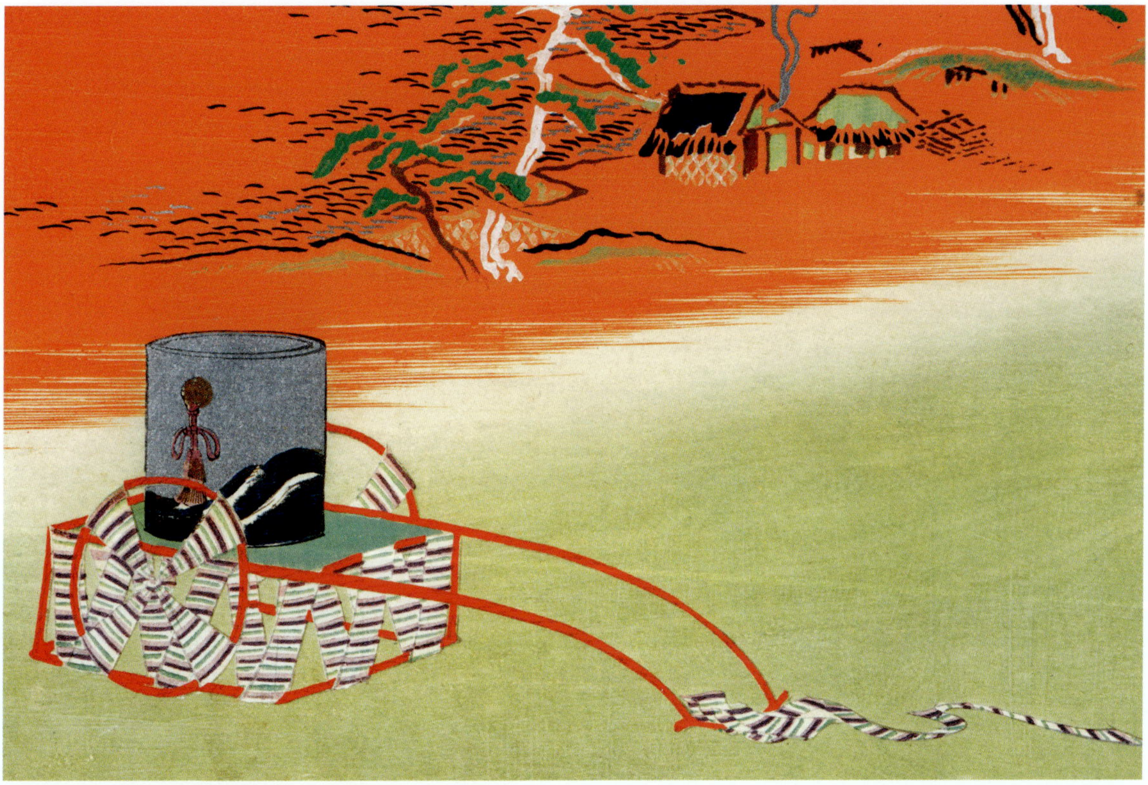

The Water Carrier with Salt Water,
by Kamisaka Sekka, aquatint, 1903
© The Art Archive/Private Collection/
Alfredo dagli Orti

But both sodium and chloride are essential for human survival. Chloride is necessary for metabolism, the chemical process that derives energy from food. Sodium regulates the volume of bodily fluids, including blood. Both sodium and chloride are available from other foods. Tomatoes, lettuce, and seaweed are sources of chloride. But the easiest way to maintain the right balance of sodium and chloride is by eating salt. This is why all mammals crave the taste—in fact, crave far more salt than needed—just as they crave sex for the biological imperative of reproducing. Nature makes us want what we need.

Another food that provides all the necessary sodium and chloride without adding salt is red meat, which is why the only societies in human history that did not make, buy, trade, or eat salt were hunter-gatherers. The history of all salt throughout the world begins with agriculture. Agricultural societies tend to have grain-based diets—rice in the case of Japan—and this diet does not supply sufficient sodium and chloride. Once livestock are being raised, they, too, require salt to eat.

Japanese history begins with hunter-gatherers, the J mon, about sixteen thousand years ago. Around 400 BCE, the J mon period ended and the people in the Japanese archipelago, near what is now Tokyo, became rice farmers. The Japanese, at that time, were undoubtedly salt producers. But there is evidence of J mon salt-making long before that. Three-thousand-year-old shards of pottery have been found in Miyagi Prefecture, in the north of Honshu Island, an area that the devastating March 2011 tsunami brought to world attention. This pottery may have been used to cook down seaweed to extract salt. There are no natural brine wells in Japan and the summers are too rainy for solar-evaporated sea salt, but Japan started to develop its own techniques. Salt could be made by passing sea water through salty sand and then boiling it in stone pans. This required a considerable amount of fuel—wood in the north, coal in the south. Solar-evaporated sea salt requires no energy, and natural brine tends to be more saline and requires less cooking. In Szechuan China, natural gas found near brine springs provided cheap energy. Once the cost of fuel was added to Japanese salt, it became an expensive product, to be used sparingly as a luxury.

Chinese contact and its subsequent enormous influence on Japan began with the Han dynasty, which dates from 206 BCE; this contact greatly accelerated the use of salt.

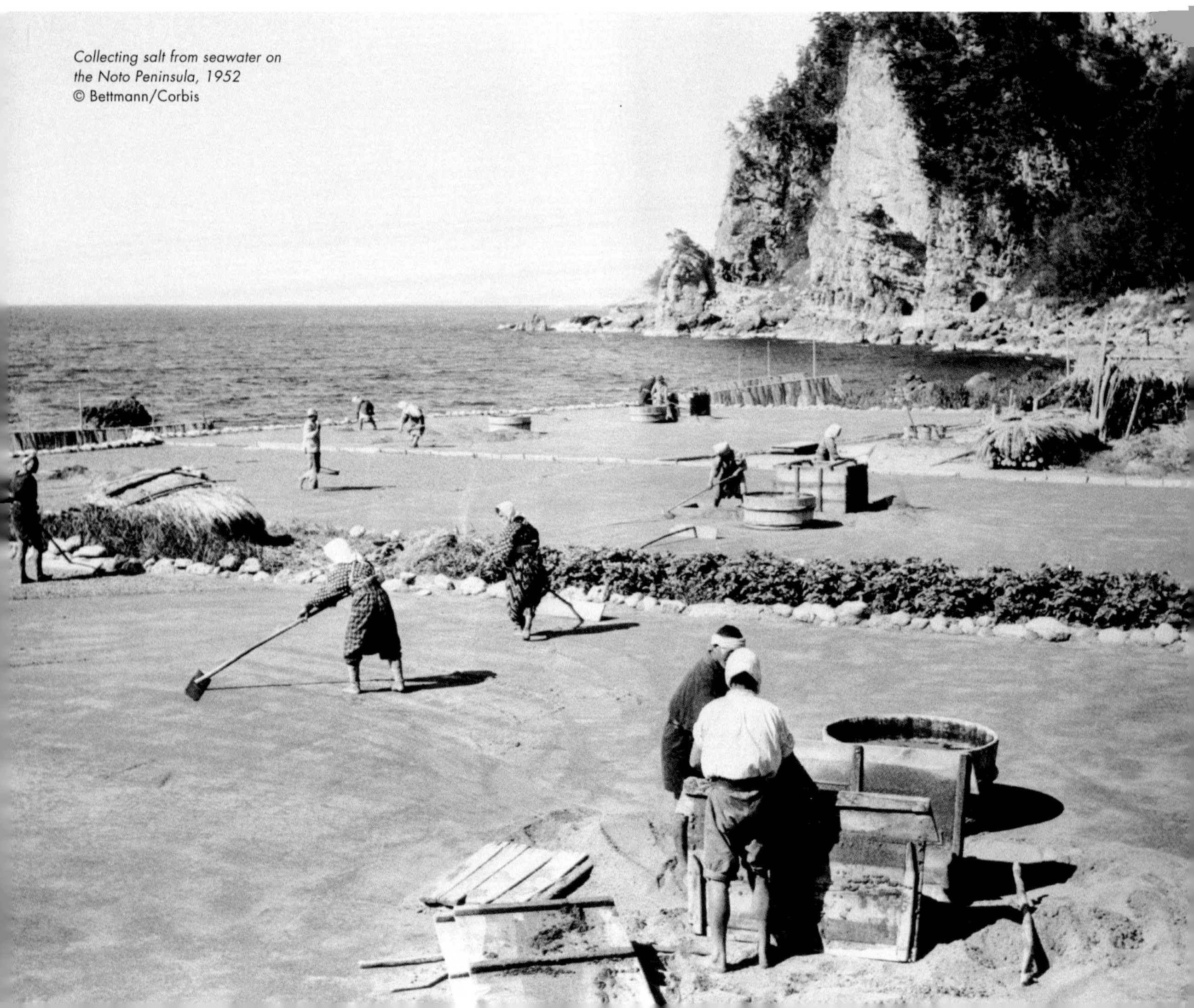

Collecting salt from seawater on the Noto Peninsula, 1952
© Bettmann/Corbis

The Chinese, at the time, were the first great salt-makers, the most technically and commercially advanced salt producers in the world. Even the word for "salt" in the Japanese language comes from Chinese. The Japanese character that means shio, salt, is adopted from the same character for "salt" in Mandarin, which is pronounced yán. The original character, in a calligraphy that went out of style with the Han, showed an imperial official boiling brine.

One way of stretching salt, learned from the Chinese, was to ferment soybeans and make a salty sauce. In Chinese cooking, salt is rarely applied directly to food; it is either pickled in brine or salted through condiments, especially soy sauce. Until the fifteenth century, soy sauce in Japan was basically a duplication of Chinese sauce, a farmhouse product made from fermenting soy and wheat in brine. But, as Japanese cooking ideas grew more sophisticated and distinct, Japanese soy sauce evolved into a commercial product. The roasted beans and wheat are exposed to mold and left to develop for three days before being liquefied with salt brine and brewed for at least a year. Japanese soy sauce has more wheat and less salt than Chinese, making it both sweeter and lighter, and Japanese cooks rely less on soy sauce for salting.

In Japanese cooking, salting is a very precise skill. Never simply shaken on food, it is gracefully sprinkled from a chef's fingers gently waving high above the food, a technique known as Furi-jio, to ensure that the salting is light and even. Fish is always salted before cooking and often before being eaten raw, as with sashimi. When grilling, the amount of salt sprinkled and the length of time it is left on the fish are determined by the size and type of fish. Fatty fish require longer and more salting than white-fleshed fish. Japanese recipes, in fact, often give the salting time.

Other foods are salted in brine, a technique called Tate-jio. The salinity of the brine is intended to resemble the ocean's, which consists of about two tablespoons to a quart of water. Fish is marinated in the brine for a maximum of a half hour, but frequently even less time. Sometimes, very fatty fish is dredged in salt—Beta-jio, meaning "salt-covered"—for an hour or longer; then, the salt is washed off just before cooking. There is also salting with a handmade paper, known as washi, made from long plant fibers. Wa means "Japanese" and shi means "paper," although the technique also originated in China. This, the most subtle salting technique, is often used for such raw food as sashimi and sushi. The washi paper is put over a salted board; the fish is placed on the paper and covered with another sheet of paper; then, salt is sprinkled on top. The salt penetrates the paper, and after about an hour

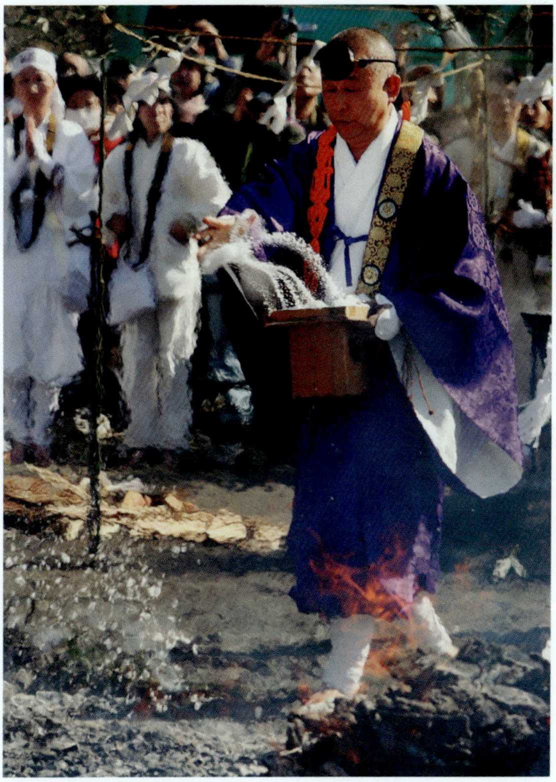

A practicing Buddhist monk throws salts to cleanse or purify the way at the start of a fire-walking ceremony, which heralds the coming of spring, Yakuoin Temple on Mt. Takao, in the city of Hachioji in western Tokyo, 2007.

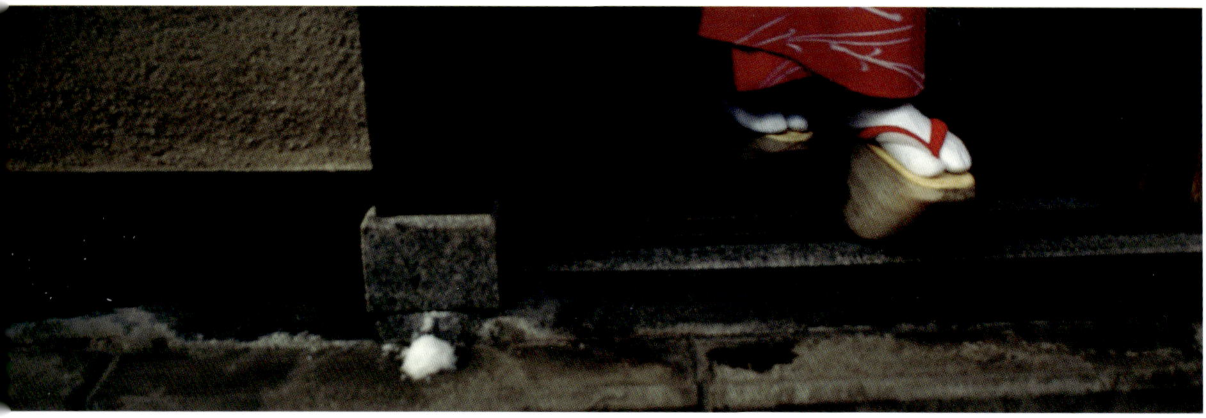

Stepping over a threshold, Kyoto
© Getty Images/
National Geographic Collection

the subtly salted fish is removed from the salt and the paper.

When whole fish are grilled, fins and tails are heavily salted so that they will not burn in grilling; this makes the fish look more presentable. Since ovens have been introduced in modern Japanese cooking, a new technique has been developed: Shio Gama, salt domes, in which the food, usually fish, is enveloped in a salty, beaten egg white and baked in this shell.

The Japanese are fish eaters. Before modern transportation, salt became particularly important to people living in the central mountains. They bought fish on the coast, salted it inside and out so thoroughly that it would last a year, and then carried it by foot on treks back to their mountain homes days away.

For most of the nation's history, salt remained a precious commodity, produced in farms along the salted beaches of Honshu. By the eighteenth century, Japan's population had grown to twenty-five million, and this increase had given rise to several cities—Osaka, Kyoto, Edo. Life was becoming less spartan. Japan was producing more salt per capita than such major western salt producers as France, but still not enough for its needs. In the nineteenth century, Japan continued to use old methods, except that iron pans had replaced stone; less fuel, therefore, was needed. By 1885, Japan produced one million tons of salt a year, but even this did not prove to be enough to fill its needs. Foreign salt was still far cheaper than the local product, so the country remained a major importer, especially from Siam and Australia. Japanese salt itself was a pure white but rather moist product that was produced throughout southern Japan, especially southwest Honshu and the northern part of the nearby island of Shikoku.

The Japanese, for much longer in their history than other countries, held onto an appreciation of the value of salt. During World War II, American soldiers fighting the Japanese became weak and ill. They had no understanding of their malady—simply the lack of sodium and chloride. In time, the U.S. Army learned to supply them with salt pills. But the Japanese army, as part of a society that still understood the importance of salt, from the outset issued every soldier, along with a rice ration, a block of salt to lick periodically.

Only in the 1970s, when Japan developed a technique of electrodialysis to concentrate sea water, did it become an important salt producer. Japan remains, though, a major salt importer.

Salt, because it is valued, continues to have symbolic importance in Japan. Much of this has to do with the ability of salt to preserve. In modern chemistry, it is understood that salt simply retards decomposition long enough for lactic acid, the real preservative, to take hold. Throughout the world, though, it is still salt that is credited with driving out rot and preserving and protecting one from evil. In Haiti, zombification is accomplished through an evil spell, and the only way to free a zombie is to use salt to drive away the evil. In many cultures, salt is brought as a gift to a new home to protect it from harm. In the Middle East, in both Jewish and Muslim cultures, the ability of salt to preserve is seen as a metaphor for keeping an agreement. A bargain—in the Jewish case, a bargain with God—can be sealed with salt. In Leonardo da Vinci's Renaissance masterpiece *The Last Supper,* a salt cellar is overturned on the table in front of Judas. If salt seals a bargain, spilling salt is a sign of a broken bargain, a betrayal. But this can be made right by tossing some of the salt over your shoulder.

Around the world, salt appears at funerals. The Irish, in earlier centuries, carried salt to wakes, taking an occasional taste, although the Hindus, a notable exception, shun salt at funerals in order to let the dead pass on.

At a traditional Japanese funeral, the body is cremated and the family lifts the bones with long chopsticks and places them in an urn, which is why food is never passed from chopstick to chopstick at a Japanese table. When the mourners return home, they purify themselves, or drive away evil, by rubbing kiyome shio, funeral salt, on their hands. Sometimes, especially in Buddhist funerals, packets of salt are distributed to the guests as they leave: They are to sprinkle the salt in their homes to keep away evil. Japanese Buddhists sometimes hold a ceremony seven days after a death, in which brine is ceremoniously sprinkled. In Japanese Buddhism, salt is tossed on a Buddha and then scooped up and placed on an injury for healing. Even today, many Japanese restaurants try to assure their success by placing mori shio, mound salt, outside the establishment at the start of the day. A small inch-and-a-half mound of salt is placed in front; sometimes two mounds are placed, sometimes three. If a mound is accidentally stepped on or kicked over, it's a good sign, a sign that the restaurant will be busy and prosperous, although polite customers usually try to avoid damaging the display. Restaurant owners usually do not push their luck by displaying the mounds too prominently. They are generally in an out-of-the-way corner.

The custom of mori shio is at least twelve centuries old in Japan—even older in China, where, it is said, the salt was used to lure animals pulling the carts of aristocrats toward the restaurant.

In sumo wrestling, salt keeps away evil. Sumo wrestling is an ancient Shinto practice, and in Japan, where so much of culture and religion comes from the Chinese, everything Shinto has a special standing because it is indigenous to Japan. The wrestling match itself, which frequently lasts less than a minute, consumes less time then the ritual preceding it. The two wrestlers step up on the sand-covered clay platform, the dohy, where the match is to take place. Each claps his hands and begins stomping his thick, powerful legs. Retreating to their respective corners, the combatants are given a ladleful of water, the chikara-mizu, to rinse their mouths, and a paper tissue, the chikara-gami, to dry their lips. Then, both step back into the ring and squat, facing each other. They clap their hands again and open them far apart, to show that they carry no weapons. But they are still not ready to wrestle. Each returns to his corner, picks up a handful of salt, and tosses it onto the ring, to drive away evil. Now, they are ready for a brief struggle. The first one of these enormous men who touches any body part to the ground—other than the soles of his feet—loses. Salt cannot guarantee victory, not even in Japan.

Mark Kurlansky has written twenty-four books, including fiction, nonfiction, biography, children's and young adult books, and has translated others. He has won numerous awards for food writing, science writing, children's books and education, and promoting peace. He has been translated into twenty-five languages. Among his several best sellers are Cod: A Biography of the Fish that Changed the World; 1968: The Year that Rocked the World; *and* Salt: A World History. Salt *first became a best seller in 2002, but recently reappeared on the* New York Times *best-seller list.*

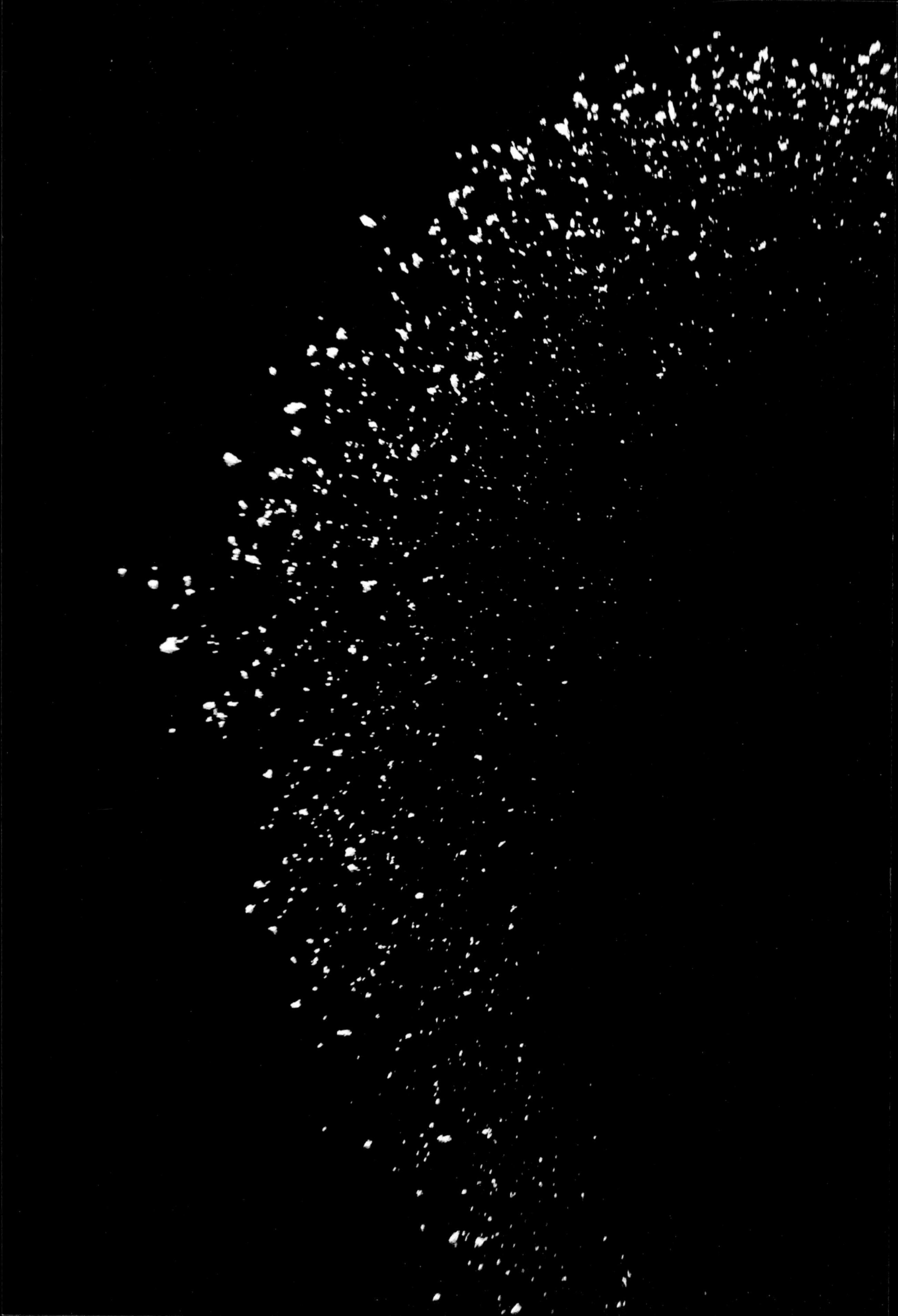

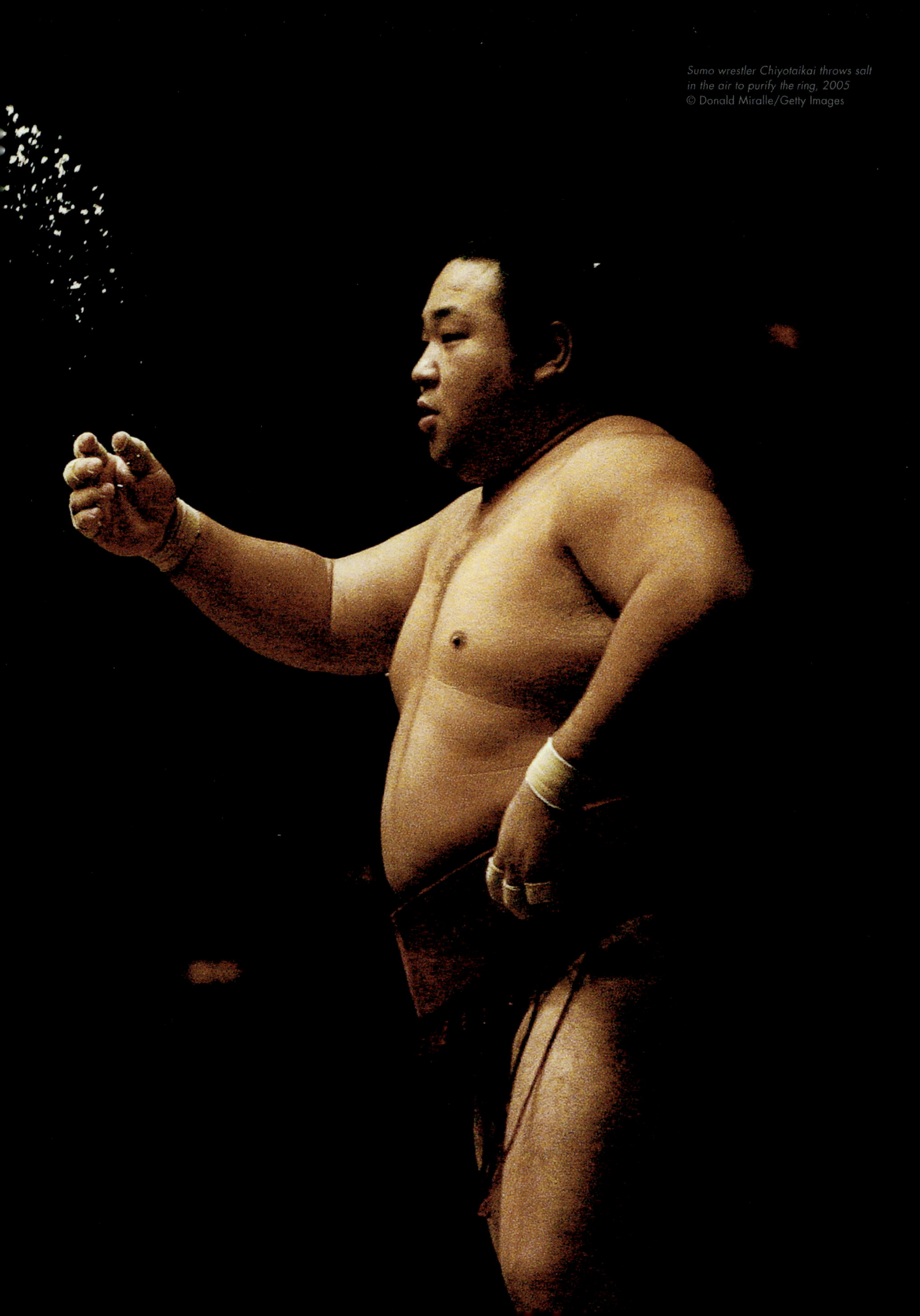

Sumo wrestler Chiyotaikai throws salt in the air to purify the ring, 2005
© Donald Miralle/Getty Images

Drawings

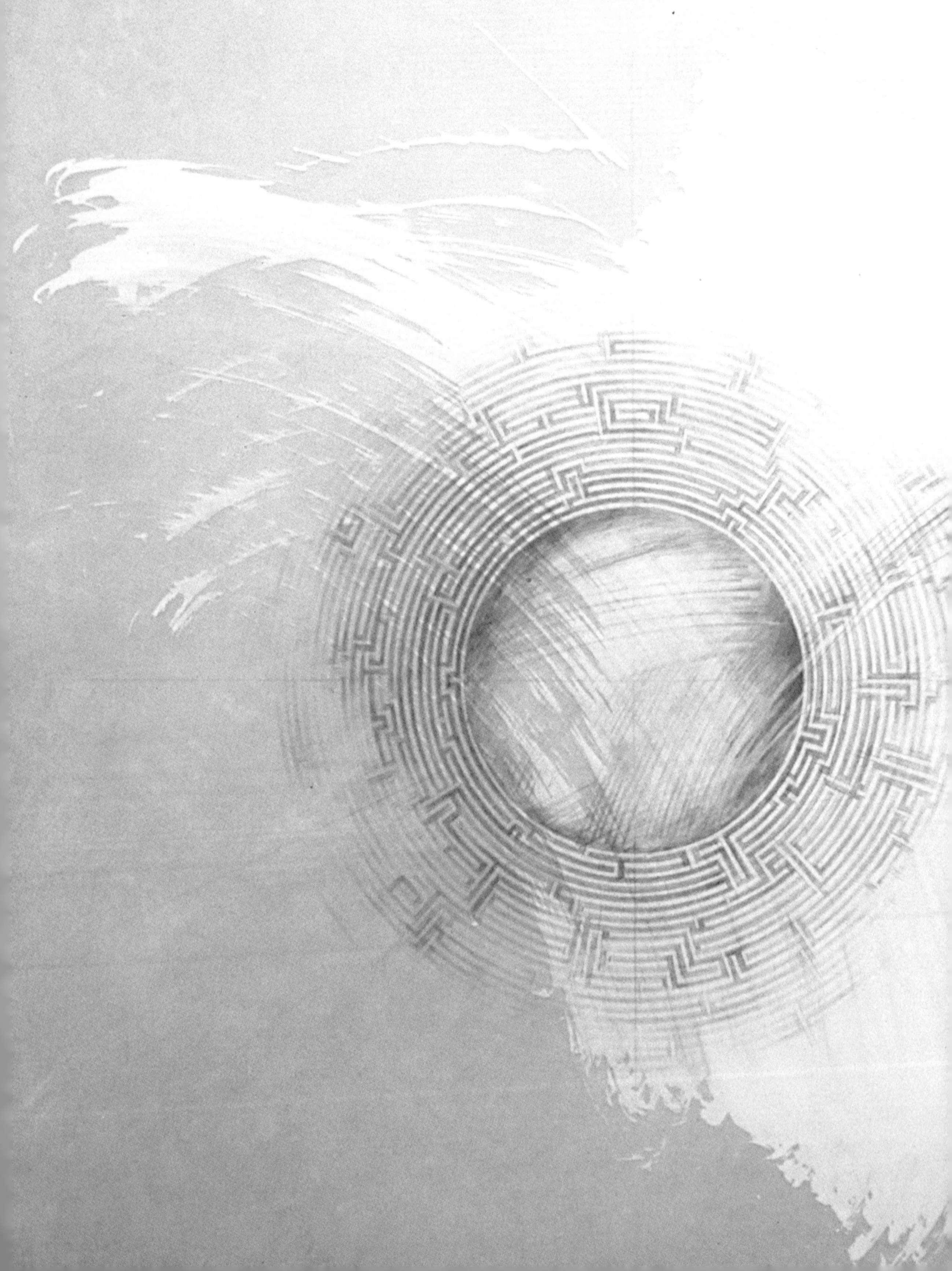

LABYRINTH
matt film, pencil, acrylic color
2001

Labyrinth

left
LABYRINTH
matt film, marker pen, pencil
2007

right
LABYRINTH
matt film, marker pen, pencil
2007

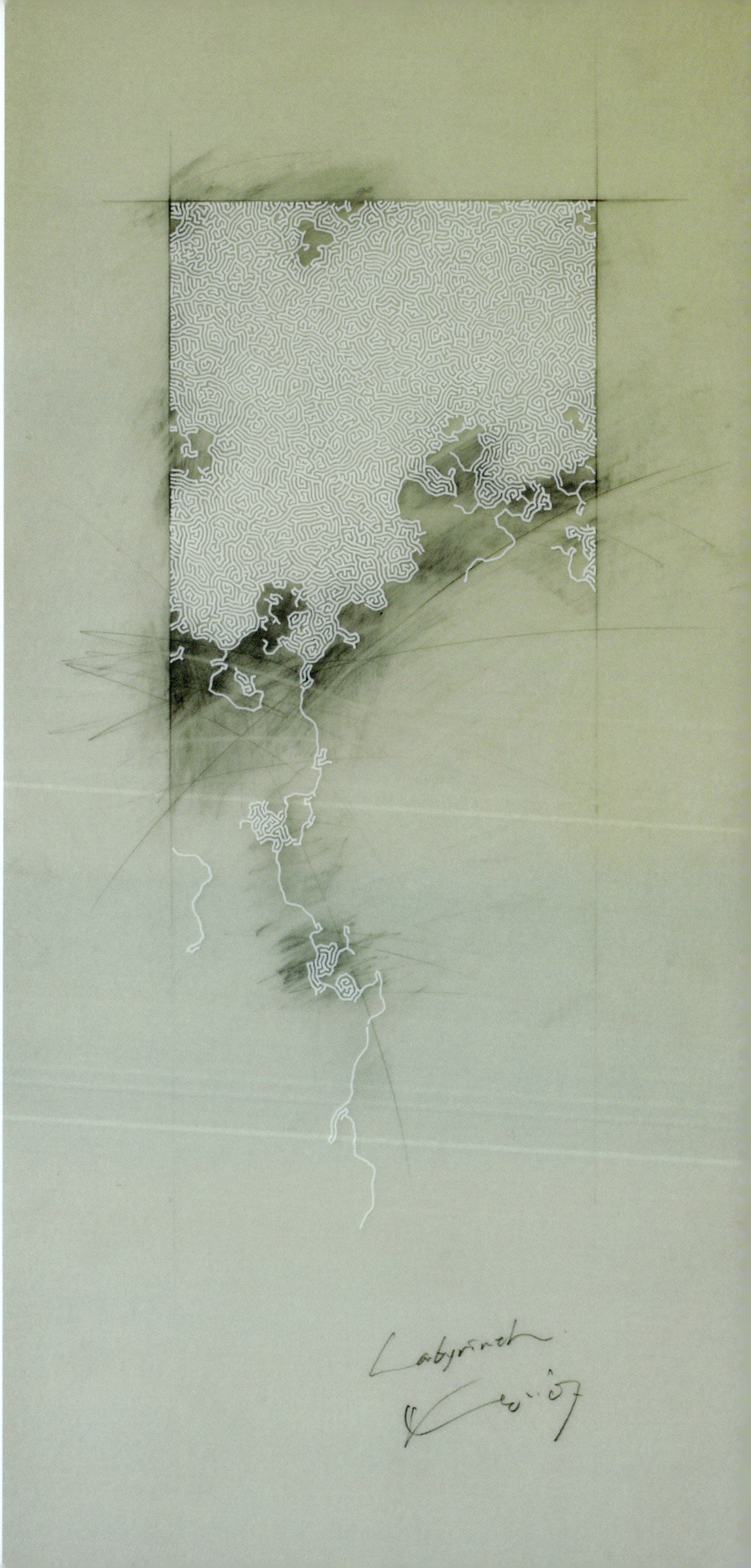

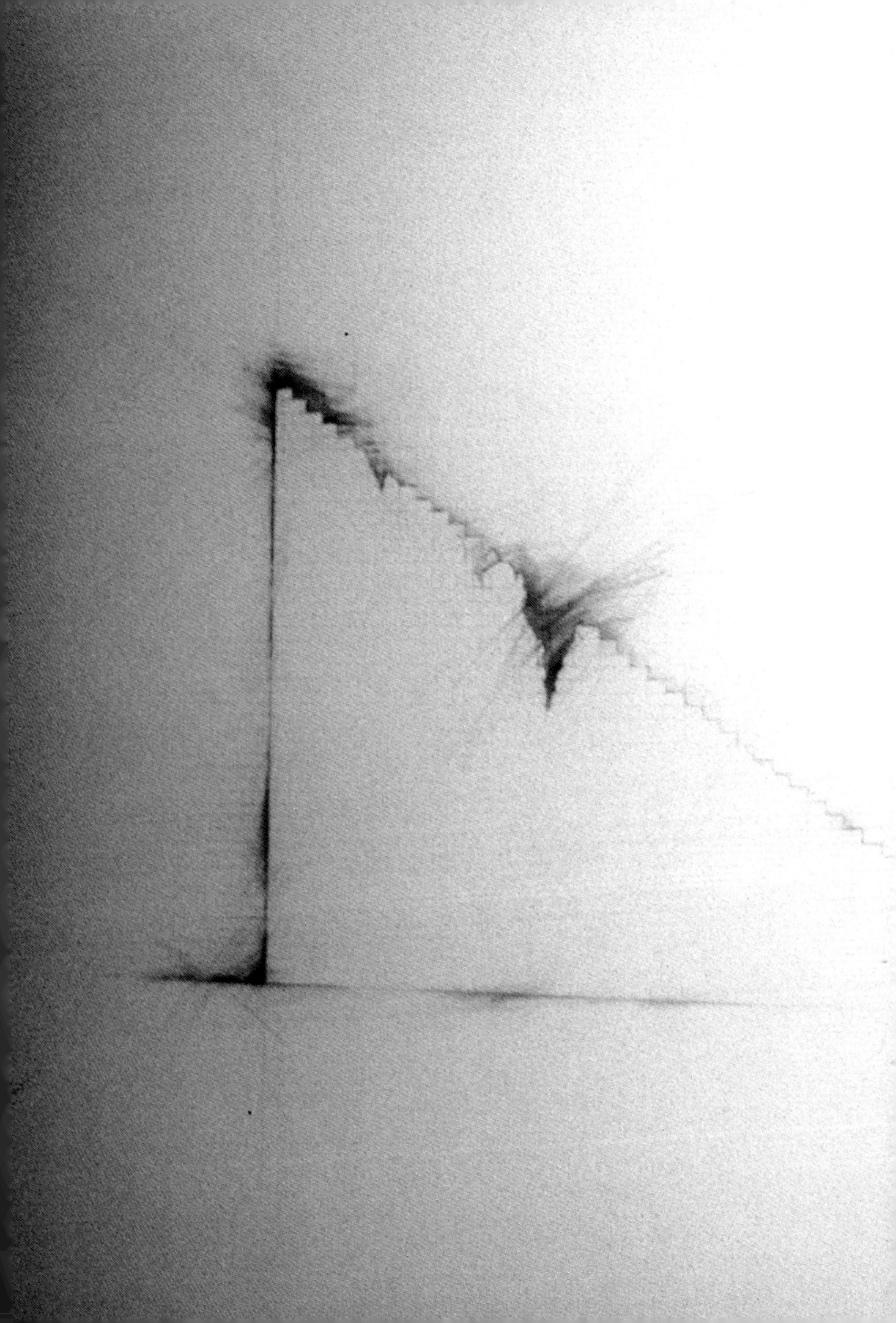

UTSUSEMI
matt film, pencil
2004

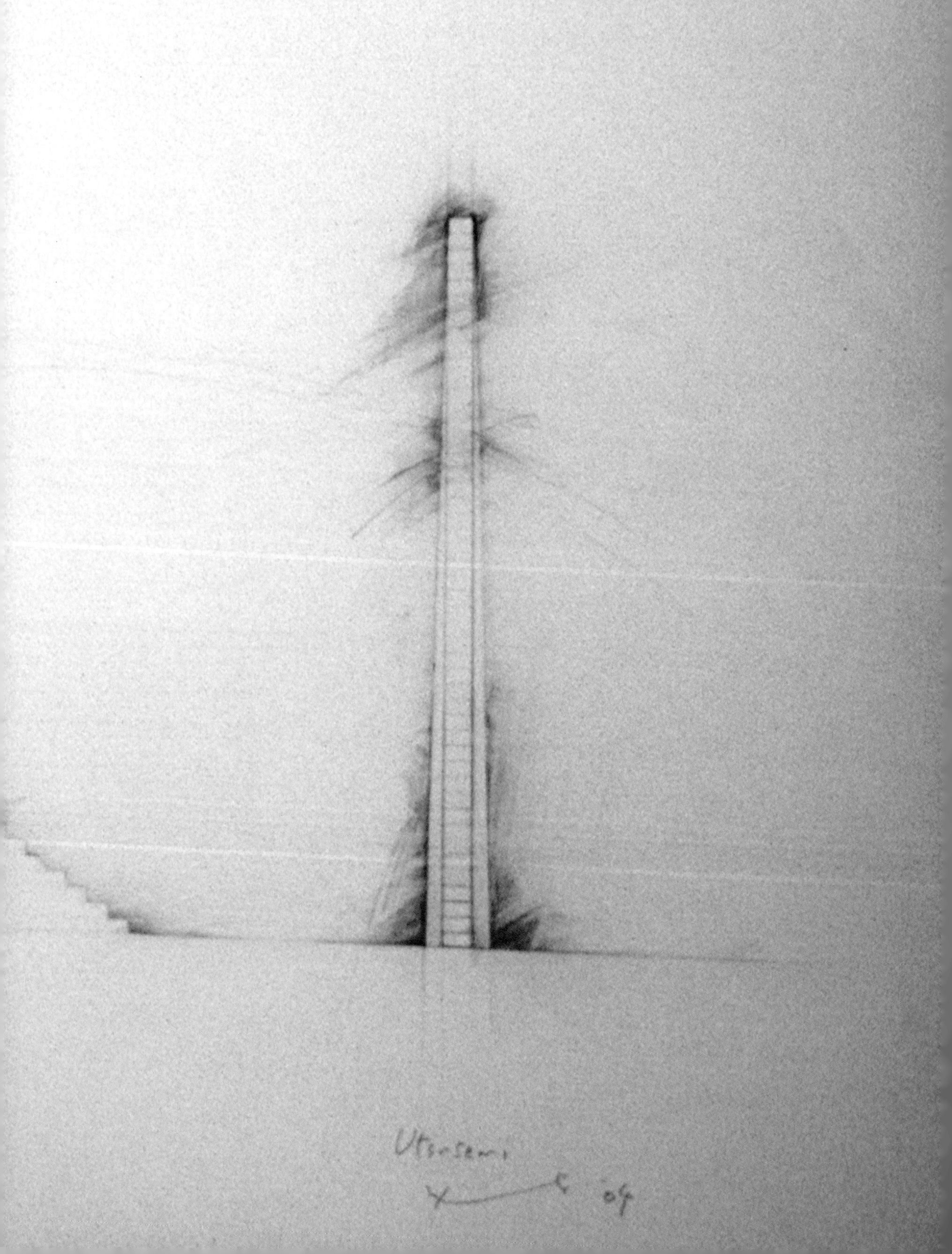

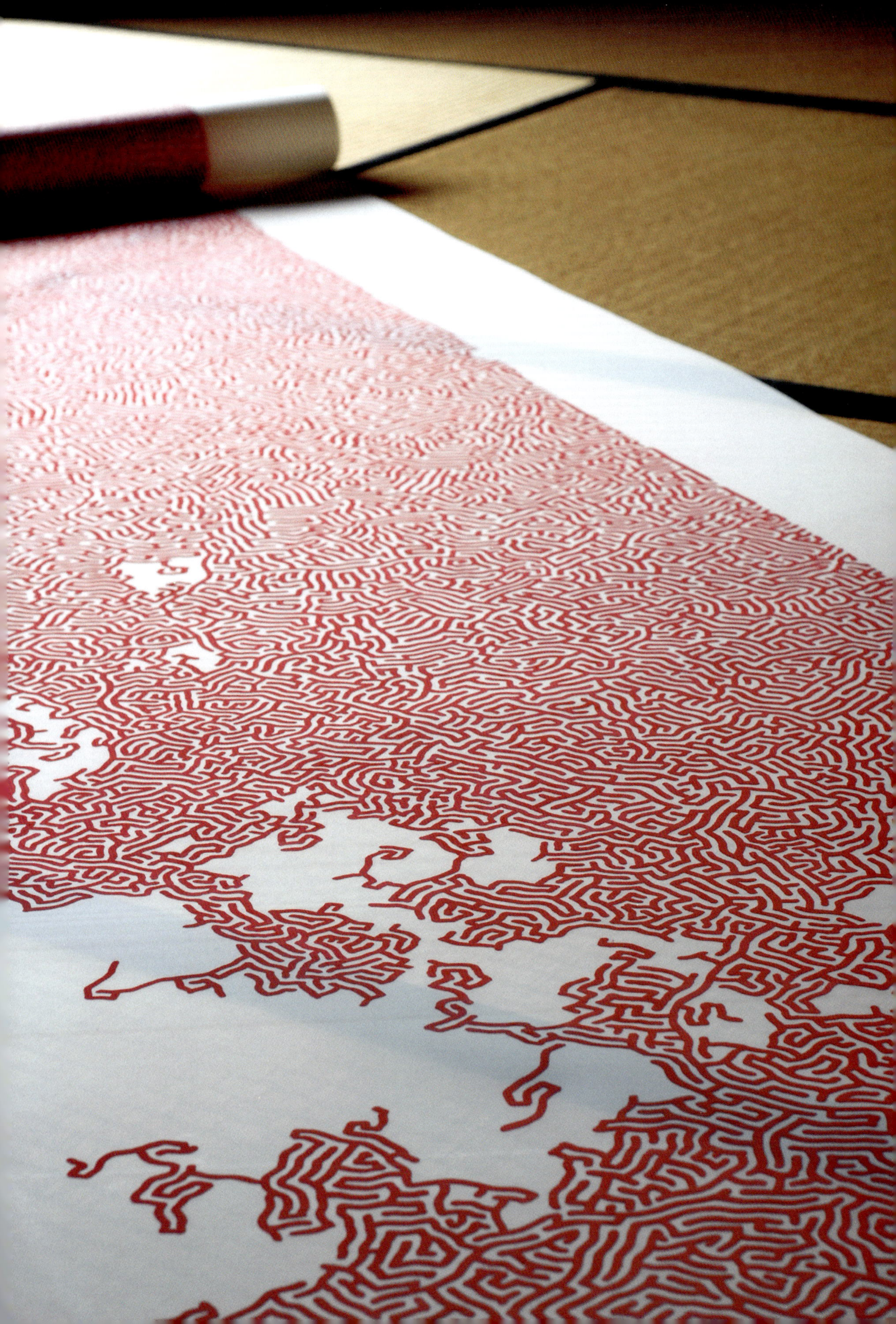

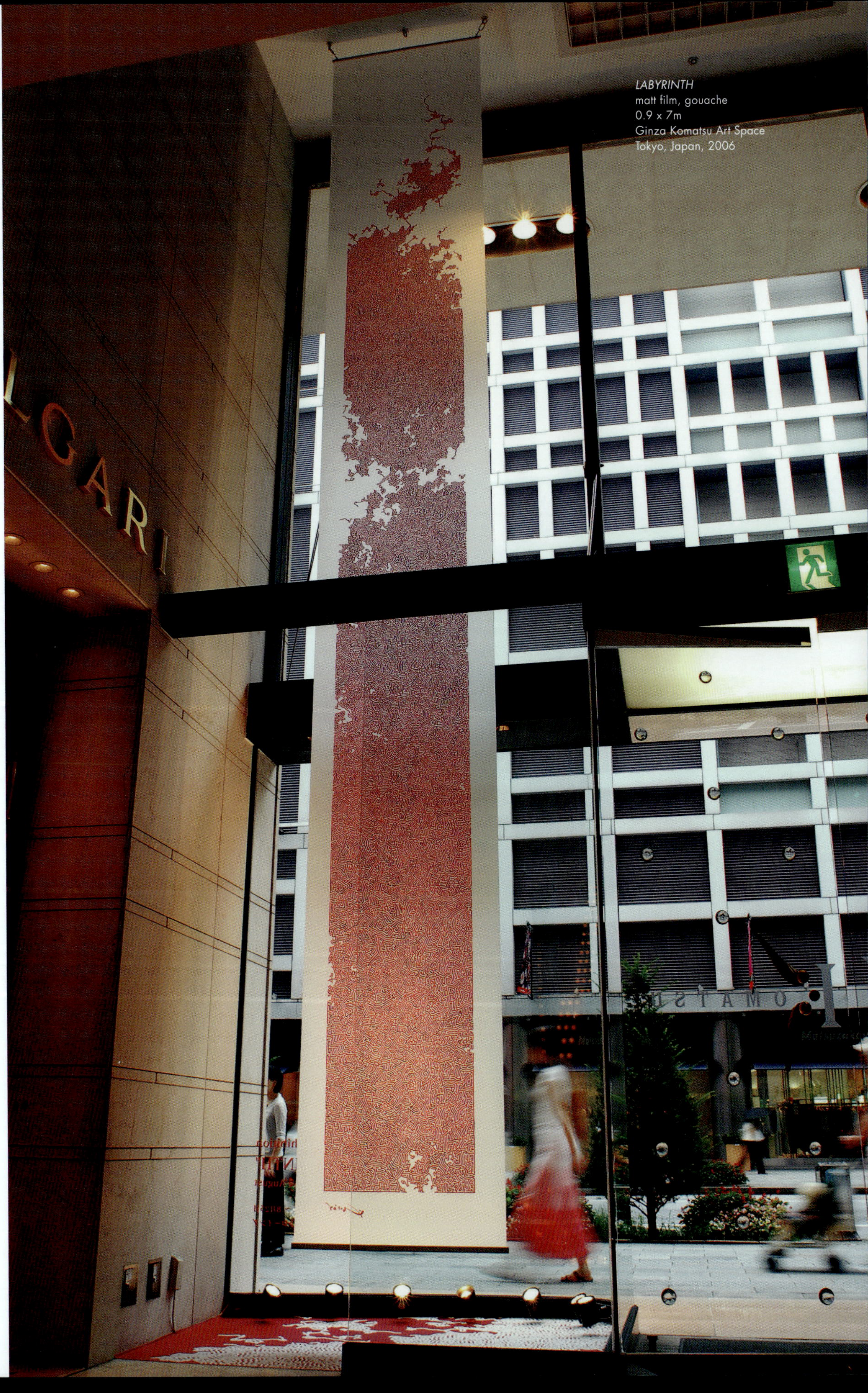

LABYRINTH
matt film, gouache
0.9 x 7m
Ginza Komatsu Art Space
Tokyo, Japan, 2006

FLOATING GARDEN
pencil, acrylic color, wood
2011

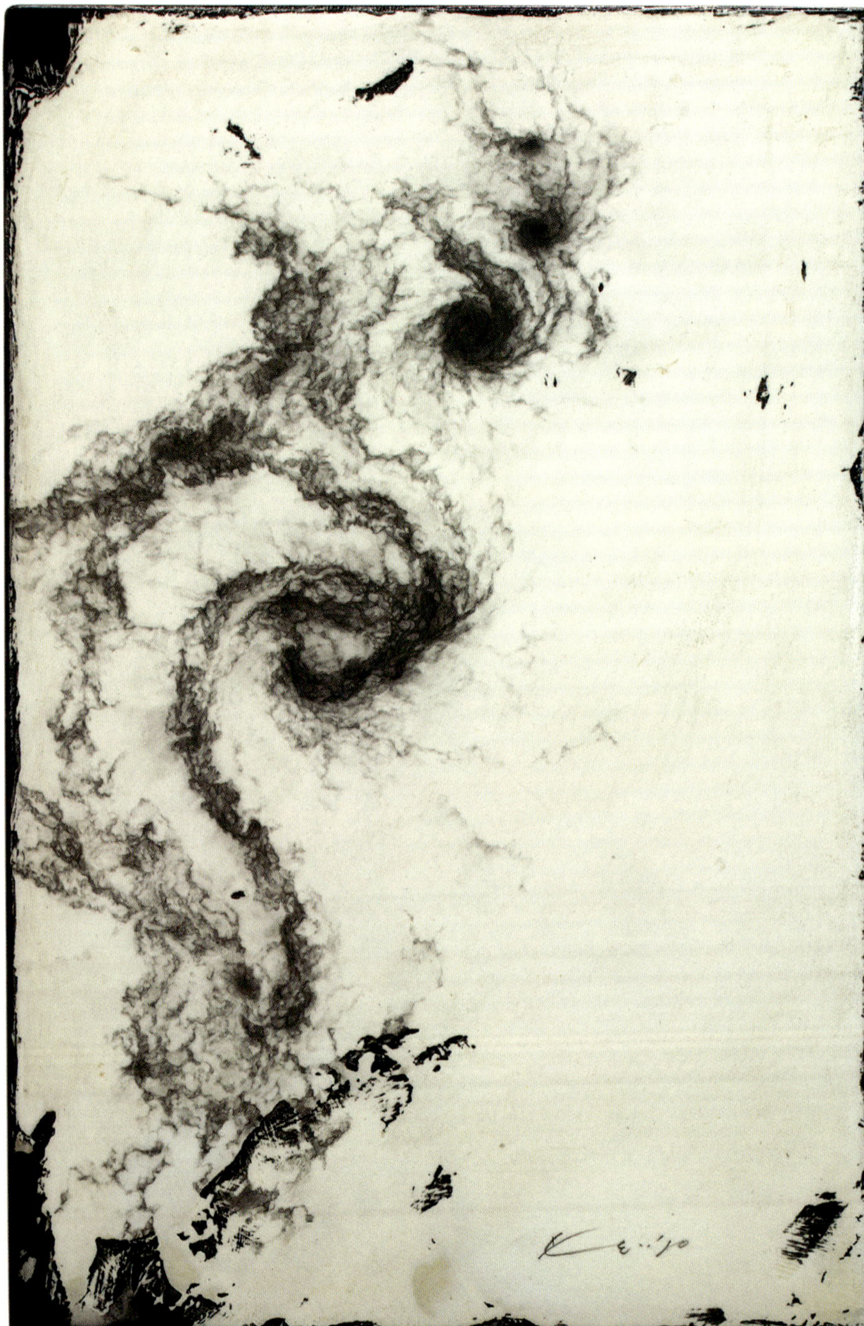

FLOATING GARDEN
negative view
giclée on Japanese paper
2011

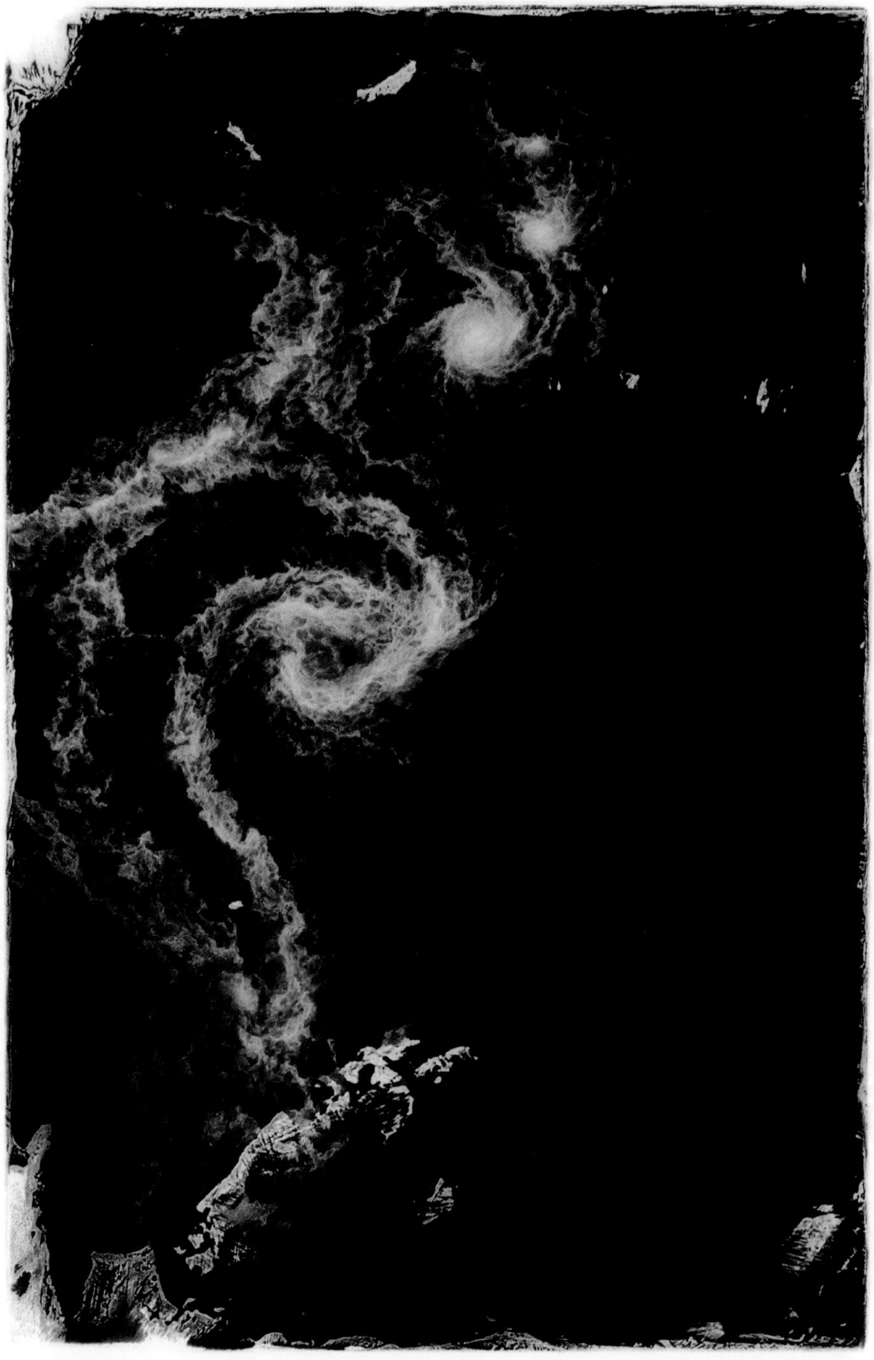

FLOATING GARDEN
pencil, acrylic color, wood
2011

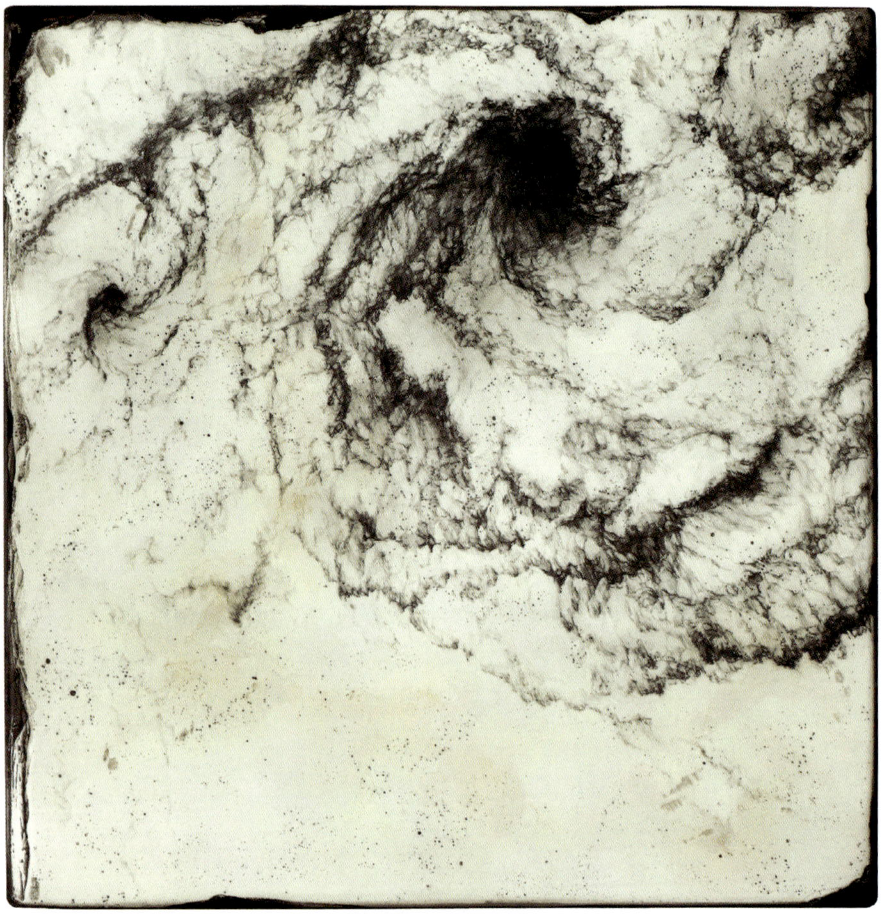

FLOATING GARDEN
negative view
giclée on Japanese paper
2011

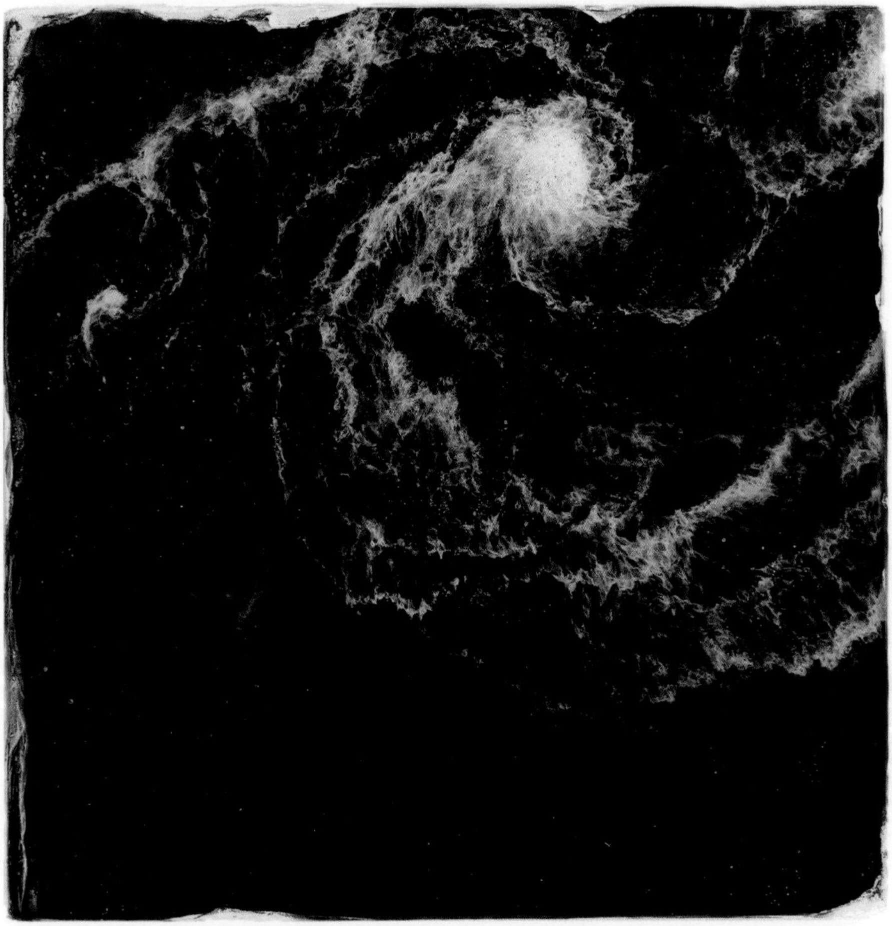

FLOATING GARDEN
pencil, acrylic color, wood
2011

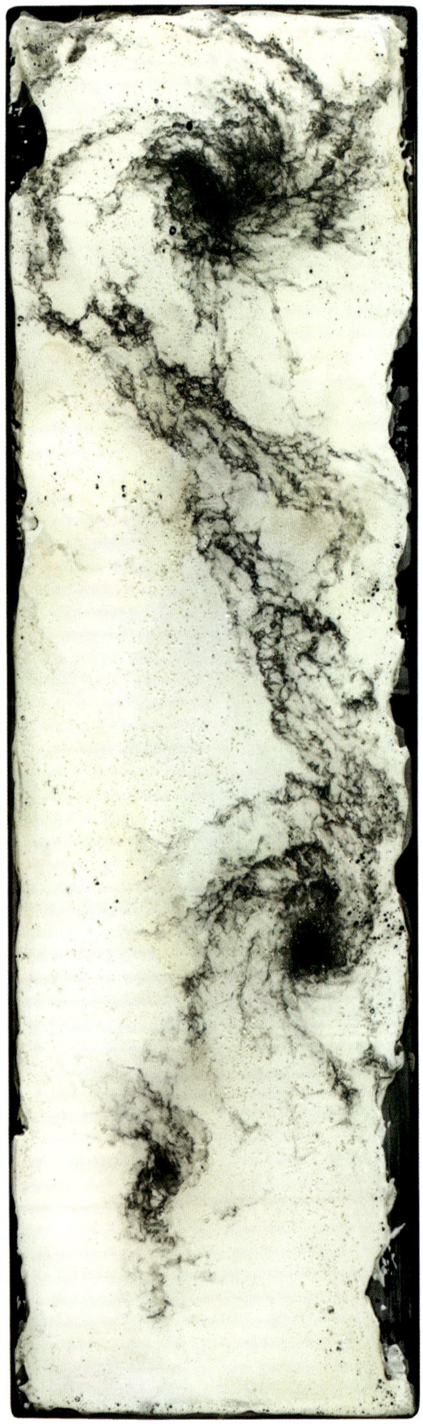

FLOATING GARDEN
negative view
giclée on Japanese paper
2011

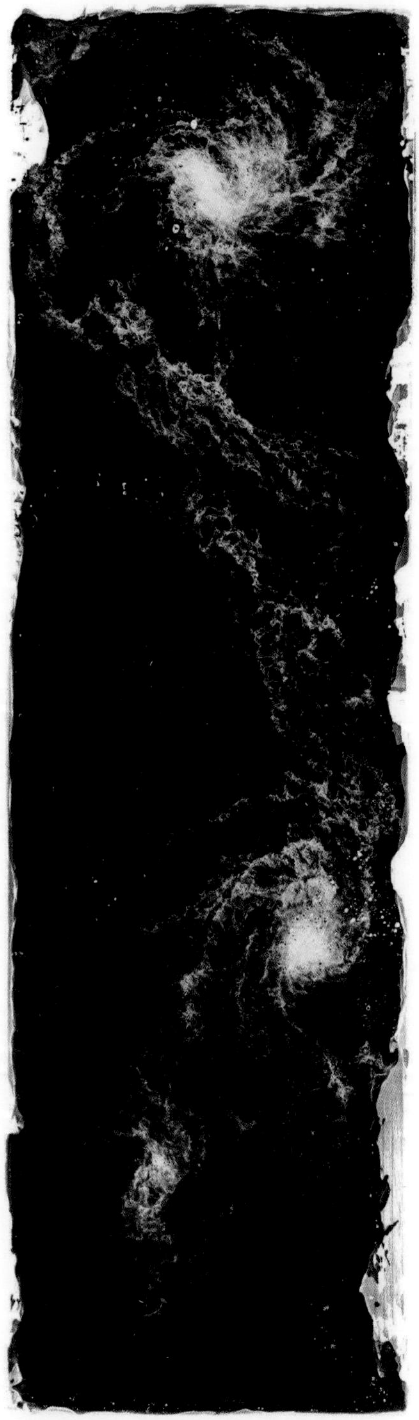

Résumé Motoi Yamamoto

1966 Born in Onomichi, Hiroshima, Japan.
1995 B.A., Kanazawa College of Art
Lives and works in Kanazawa, Ishikawa, Japan.
2002 Philip Morris Art Award
2003 Recipient Pollock-Krasner Foundation, Inc., grant, New York
2010 Recipient Voyager + AIT scholarship program grant, Tokyo

Selected Solo Exhibitions

2011 The Hakone Open-Air Museum, Kanagawa, Japan
2010 *Art-O-Rama*, L MD Gallery, Marseilles, France
Kunst Station St. Peter, Cologne, Germany
eN-arts, Kyoto, Japan
2009 L MD Gallery, Paris, France
Grosse Bleichen Bilder, Hamburg, Germany
Mikiko Sato Gallery, Hamburg, Germany
2006 Ginza Komatsu Art Space, Tokyo, Japan
2005 Ierimonti Gallery, Milan, Italy
Nizayama Forest Art Museum, Toyama, Japan
Art Gallery Artium, Fukuoka, Japan
CAI, Contemporary Art International, Hamburg, Germany
2002 T.L.A.P., Tokyo, Japan
2001 Akiyama Gallery, Tokyo, Japan
Gallery K2, Ishikawa, Japan
2000 Kojimachi Gallery, Tokyo, Japan
300 days Gallery, Tokyo, Japan
Gallery G2, Fukui, Japan
1999 Gallery Bellini Yokohama, Kanagawa, Japan
Gallery LeDeco, Tokyo, Japan
1998 Gallery Moe, Tokyo, Japan
Gallery Arai, Shizuoka, Japan
1997 Gallery Ando, Tokyo, Japan
Gallery Rasen, Toyama, Japan
1996 Gallery Myu, Tokyo, Japan
1994 Kanazawa Yomiuri Hall, Ishikawa, Japan

Selected Group Exhibitions

2012 *Making Mends,* Bellevue Arts Museum, Washington
2011 *City-Net Asia 2011*, Seoul Museum of Art, Seoul, Korea
Reliefs, Fondation espace ecureuil pour l'art Contemporain, Toulouse, France
2010 *MOT ANNUAL 2010, Neo-Ornamentalism from Japanese Contemporary Art*, Museum of Contemporary Art, Tokyo, Japan
2009 *Machiya-Junyu, Konichi*— Soy Sauce Factory, Ishikawa, Japan
Hundred Stories About Love, 21st Century Museum of Contemporary Art, Kanazawa, Ishikawa, Japan
Togo Murano x Motoi Yamamoto, Kanazawa Art Gummi, Ishikawa, Japan
Tanagokoro 9, Radium Roentgen Werke, Tokyo, Japan
2008 *Conceptions Taken from, Funa-Asobi*, Ishikawa, Japan
Landschaft, Radium Roentgen Werke Tokyo, Japan
Text of Life, Ashikaga Museum of Art, Tochigi, Japan
Black, White and Gray, MA2 Gallery, Tokyo, Japan
2007 *AIR Onomichi 2007,* Hiroshima, Japan
Force of Nature, part 2, Sumter County Gallery of Art, South Carolina

2006 *Force of Nature*, Van Every/Smith Gallery, Davidson College, North Carolina
Force of Nature, Halsey Institute of Contemporary Art, Charleston, South Carolina
C.A.R.K. 2006, Ishikawa International Salon, Ishikawa, Japan
C.A.R.K. 2006, Maragopoulos, Patras, Greece
The Library, Ashikaga Museum of Art, Tochigi & Tama Art University Museum, Toyko, Japan

2005 *Rising Sun, Melting Moon—Contemporary Art from Japan,* The Israel Museum, Jerusalem, Israel
hoch hinaus, Kunstmuseum, Thun, Switzerland
The Road Not Taken '05 Kyoto Gallery Sowaka, Kyoto, Japan

2004 *The Encounters in the 21st Century, Polyphony—Emerging Resonances,* 21st Century Museum of Contemporary Art, Kanazawa, Ishikawa, Japan
Gwangju Biennale 2004, "Eco-Metoro Project," Korea

2003 *The First Steps, Emerging Artist from Japan*, P.S.1, Contemporary Art Center, New York
Landschaft, Radium Roentgen Werke, Tokyo, Japan
Mutated Zen, The Nunnery, London, UK
Respiri, Galleria Vittorio Emanuele, Milan, Italy

2002 *Philip Morris Art Award 2002,* Tokyo, Japan
Corresponding Sites: Four Spaces—Four Artists, Hamburg, Germany

2000 *Philip Morris Art Award 2000,* Ebisu Garden Place, Tokyo, Japan
Installation, Garden of the Sculptures, Veracruz, Mexico

1999 *The World Tour*, Kanazawa Citizens' Art Center, Ishikawa, Japan
Standing in the Future, Telecom-center, Tokyo, Japan

1998 *Pre-Opening Event, 21st Century Museum of Contemporary Art, Kanazawa,* Ishikawa, Japan
The Library Kanazawa, Kanazawa Citizens' Art Center, Ishikawa, Japan

1997 *Hiroshima Art Competition*, Hiroshima City Museum of Contemporary Art, Hirosima, Japan
Art Exchange Society '97('98), Shinkiba Soko Gallery, Tokyo, Japan

1996 *Representation of Contemporary Art,* Kanazawa Citizens' Art Center, Ishikawa, Japan

1995 *Art Festival in Tsurugi '95 ('96–'98),* Ishikawa, Japan

1994 *The 20th France-Japan Contemporary Art Exhibition*, L- Prize Tokyo & Ishikawa, Japan
International Competition Tannan Art Festival '94 ('95–'01), Fukui, Japan

1992 *The 31st Hokuriku Chunichi Art Exhibition,* Grand Prize, Ishikawa, Japan

Photo: em yamaguchi

Curator's Acknowledgments

This project began with a conversation on the shores of Lake Eden, on the site of the former Black Mountain College in North Carolina in 2002. My friend and colleague Brad Thomas and I were discussing the relationship between Zen Buddhism and modern art, especially as expressed through the work of Black Mountaineer John Cage. His musical compositions based on the placement of knots in the bark of a tree, for instance, or his reliance on "chance operations" to generate form were some of the topics of the discussion. On the spot, we vowed to get to the bottom of this vexing question by planning a trip to Japan to find and identify the origins of the creative impulse that wishes to work with nature to create art. In 2003, with the help of a grant from the Freeman Foundation, we found ourselves bound for an unforgettable curatorial foray throughout Japan, interviewing artists known to be inspired by natural materials or natural processes in the very act of creation. We sorted through the work of more than three hundred artists who were identified for us—then interviewed over thirty artists, to narrow the list down to ten—Motoi Yamamoto being among them.

These ten artists were brought to the United States for simultaneous six-week residencies, creating work for the exhibition *Force of Nature: Site Installations by Ten Japanese Artists* in the fall of 2006. Brad Thomas and I were invited back to Japan in the summer of 2007 for a symposium on the theme hosted by Kyoto University of Art and Design and to give a series of lectures on the project in Osaka and Kanazawa. The entire *Force of Nature* experience has proven one of the most rewarding of my professional life.

I give special thanks to Brad Thomas for being game to take on a project of such ambition and magnitude back in 2002. Our curatorial partner, guide, translator, and esteemed colleague at Kyoto University of Art and Design, Noriko Fuku, deserves special gratitude. Ken Lambla, dean of the School of Arts + Architecture at the University of North Carolina, Charlotte; Robert Miller, (former) director of the Clemson Architecture Center in Charleston; Tom Stanley, chair of fine arts at Winthrop University; Karen Watson, director of the Sumter Gallery of Art; and Ce Scott, (former) creative director of the McColl Center for Visual Art, were all able and committed partners in this project. Buff Ross, web and catalogue designer; Mitchell Kearney, project photographer; and Halsey staffers Katie Lee and Katherine Borkowski all deserve a tip of the hat.

For the current project, I am indebted to David Pastre and Ray Huff, Clemson Architecture Center in Charleston; Gil Shuler, graphic designer; Halsey staffers Karen Ann Myers, Tatjana Beylotte, Lizz Biswell, and Lisa Reed; College of Charleston administrators dean Valerie Morris, provost George Hynd, and Steve Osborne; Asian studies chair Mary Beth Heston; and translator Miyako Fujiwara.

Return to the Sea: Saltworks by Motoi Yamamoto is funded in part by a grant from the Japan-United States Arts Fellowship Program of the Asian Cultural Council. Additional funding provided by the National Endowment for the Arts and the E. Rhodes and Leona B. Carpenter Foundation.

asian cultural council

ART WORKS.
arts.gov

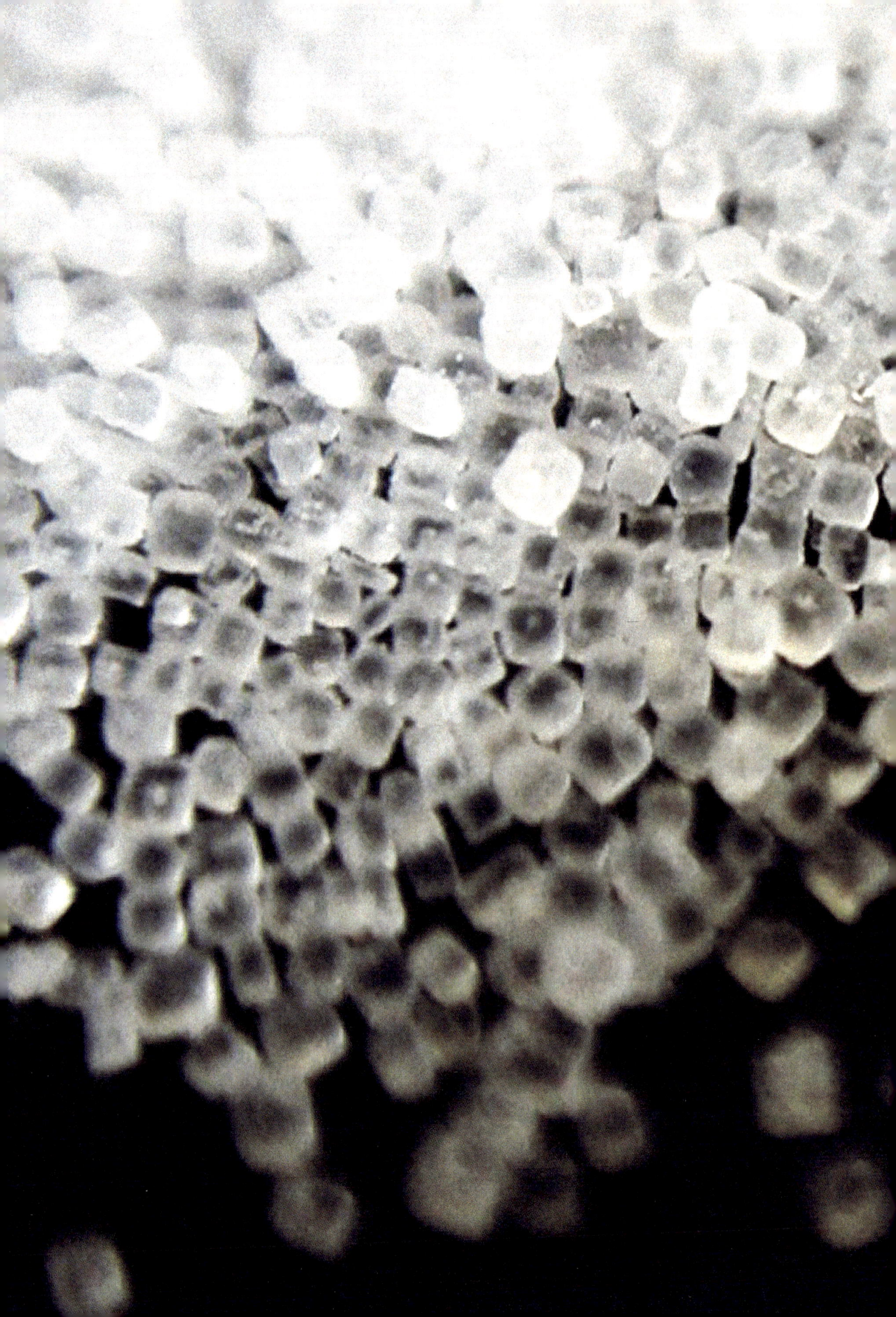